Honda 750 & 900 dohc Fours Owners Workshop Manual

by Pete Shoemark

Models covered

CB750 K. 749cc. Introduced UK October 1978, US September 1979
CB750 K LTD. 749cc. Introduced US only January 1979
CB750 F. 749cc. Introduced UK February 1980, US January 1979
CB750 F2. 749cc. Introduced UK only February 1982
CB750 C. 749cc. Introduced US only October 1979
CB750 SC. 749cc. Introduced US only January 1982
CB900 F. 901cc. Introduced UK January 1979, US September 1981
CB900 F2. 901cc. Introduced UK only March 1981

ISBN 978 1 85010 217 5

© Haynes Publishing Group 1993

ABCDE
FGHIJ
KL

3

Printed in Malaysia   (535–5S11)

British Library Cataloguing in Publication Data
A catalogue record for this book is available from the British Library

Library of Congress Control Number 86-80032

THE BOOK

Haynes Publishing Group
Sparkford, Nr Yeovil
Somerset BA22 7JJ, England

Haynes North America, Inc
859 Lawrence Drive
Newbury Park
California 91320, USA

Acknowledgements

Our thanks are due to Paul Branson Motorcycles of Yeovil and P.R. Taylor and Sons of Calne, who loaned the machines featured in the photographs throughout this manual. Alan Jackson assisted with the strip-down and rebuilding of the machine, and devised many of the ingenious methods to overcome the lack of service tools. Tony Stedman arranged and took the photographs which accompany the text. Mansur Darlington edited the text.

Finally, we would also like to thank the Avon Rubber Company who provided information on tyre fitting, and NGK Spark Plugs (UK) Limited, who furnished advice about sparking plug conditions.

About this manual

The purpose of this manual is to present the owner with a concise and graphic guide which will enable him to tackle any operation from basic routine maintenance to a major overhaul. It has been assumed that any work would be undertaken without the luxury of a well-equipped workshop and a range of manufacturer's service tools.

To this end, the machine featured in the manual was stripped and rebuilt in our own workshop, by a team comprising a mechanic, a photographer and the author. The resulting photographic sequence depicts events as they took place, the hands shown being those of the author and the mechanic.

The use of specialised, and expensive, service tools was avoided unless their use was considered to be essential due to risk of breakage or injury. There is usually some way of improvising a method of removing a stubborn component, provided that a suitable degree of care is exercised.

The author learnt his motorcycle mechanics over a number of years, faced with the same difficulties and using similar facilities to those encountered by most owners. It is hoped that this practical experience can be passed on through the pages of this manual.

Where possible, a well-used example of the machine is chosen for the workshop project, as this highlights any areas which might be particularly prone to giving rise to problems. In this way, any such difficulties are encountered and resolved before the text is written, and the techniques used to deal with them can be incorporated in the relevant sections. Armed with a working knowledge of the machine, the author undertakes a considerable amount of research in order that the maximum amount of data can be included in this manual.

Each Chapter is divided into numbered sections. Within these sections are numbered paragraphs. Cross reference throughout the manual is quite straightforward and logical. When reference is made 'See Section 6.10' it means Section 6, paragraph 10 in the same Chapter. If another Chapter were intended the reference would read, for example, 'See Chapter 2, Section 6.10'. All the photographs are captioned with a section/paragraph number to which they refer and are relevant to the Chapter text adjacent.

Figures (usually line illustrations) appear in a logical but numerical order, within a given Chapter. Fig. 1.1 therefore refers to the first figure in Chapter 1.

Left-hand and right-hand descriptions of the machines and their components refer to the left and right of a given machine when the rider is seated normally.

Motorcycle manufacturers continually make changes to specifications and recommendations, and these, when notified, are incorporated into our manuals at the earliest opportunity.

We take great pride in the accuracy of information given in this manual, but motorcycle manufacturers make alterations and design changes during the production run of a particular motorcycle of which they do not inform us. No liability can be accepted by the authors or publishers for loss, damage or injury caused by any errors in, or omissions from, the information given.

Contents

Left-hand view of the Honda CB750K

Right-hand view of the Honda CB900F

Close-up of engine/gearbox unit of the Honda CB750K

Introduction to the Honda DOHC Fours

Although for many years manufacturers have produced what are virtually replicas of the 'Works' racing machines, for sale to the general public, it is true to say that no machine captured the interest of the motor cyclist so vividly as the first production version of the Honda 4 cylinder. Already a legend in racing circles, the Honda 4 had represented a serious challenge whenever it appeared in International events. With riders such as the late Bob MacIntyre, Jim Redman and Mike Hailwood, the 4 demonstrated its supremacy on frequent occasions, irrespective of whether the 125 cc, 250 cc, 350 cc or 500 cc version was raced. Even the previously unbeaten multi-cylinder Italian models no longer had things their own way and were hard put to continue racing under truly competitive terms.

At the end of 1967 Honda withdrew from racing and commenced work on a road-going version of their in-line 4, scaled up to 750 cc. Without question it was designed to be the number one 'Superbike'. In engine layout it followed the lines of the racing machines closely, a feature heightened by the use of four separate carburettors and four sets of exhaust pipes and silencers, two on each side of the machine. A speedometer calibrated up to 150 mph and a tachometer with the red band commencing at 8500 rpm completed the 'street race' effect, which led to such a peak of interest that over 61,000 Honda 750 cc 4's were sold **in the USA alone** in just over three years.

In Britain the 750 model was first imported during January 1970, designated the model CB750.

After a production run of almost ten years, the ubiquitous Honda four cylinder models had begun to show their age when compared with the dohc four cylinder models offered by the other Japanese manufacturers. This prompted the introduction of the new line of Honda fours, designed to take the configuration into the 1980s.

A completely new engine gearbox unit is fitted, featuring a dohc (double overhead camshaft), four-valve head. In addition, there are numerous detail refinements and a complete re-styling job.

The basic model of the range is the CB750K, a touring roadster version with the traditional four-into-four exhaust system. The CB750F is the sports version and reflects the popular 'Euro-style' integrated fuel tank, side panel and seat assembly. A less bulky four-into-two exhaust system adds to the image and subtracts from the weight. In Europe only, there is the CB900F, a larger capacity version of the above model, an identical chassis housing the more powerful engine.

The range is completed by the CB750K LTD (Limited Edition) and CB750C Custom models, each reflecting the current popularity of factory 'customised' motorcycles. These two models are similar in all major respects and are treated together in the text.

Model dimensions and weights

Overall length	CB750K:	2200 mm (87.4 in)
	CB750K LTD:	2290 mm (90.2 in)
	CB750C:	2300 mm (90.6 in)
	CB750F:	2195 mm (86.4 in)
	CB900F:	2240 mm (88.2 in)
Overall width	CB750K:	880 mm (34.6 in)
	CB750K LTD:	880 mm (34.6 in)
	CB750C:	919 mm (36.2 in)
	CB750F:	865 mm (34.1 in)
	CB900F:	795 mm (31.3 in)
Overall height	CB750K:	1160 mm (45.7 in)
	CB750K LTD:	1160 mm (45.7 in)
	CB750C:	1165 mm (45.9 in)
	CB750F:	1140 mm (44.9 in)
	CB900F:	1125 mm (44.3 in)
Wheelbase	CB750K:	1520 mm (59.8 in)
	CB750K LTD:	1520 mm (59.8 in)
	CB750C:	1526 mm (60.1 in)
	CB750F:	1520 mm (59.8 in)
	CB900F:	1515 mm (59.6 in)
Seat height	CB750K:	800 mm (31.5 in)
	CB750K LTD:	785 mm (30.9 in)
	CB750C:	759 mm (29.9 in)
	CB750F:	810 mm (31.9 in)
	CB900F:	815 mm (32.1 in)
Ground clearance	CB750K:	150 mm (5.9 in)
	CB750K LTD:	145 mm (5.7 in)
	CB750C:	129 mm (5.1 in)
	CB750F:	140 mm (5.5 in)
	CB900F:	150 mm (5.9 in)
Dry weight	CB750K:	233 kg (512 lb)
	CB750K LTD:	234 kg (516 lb)
	CB750C:	232 kg (511 lb)
	CB750F:	230 kg (507 lb)
	CB900F:	233 kg (512 lb)

Ordering Spare Parts

When ordering spare parts for any Honda model it is advisable to deal direct with an official Honda agent, who should be able to supply most items ex-stock. Parts cannot be obtained from Honda (UK) Limited direct: all orders must be routed via an approved agent, even if the parts required are not held in stock.

Always quote the engine and frame numbers in full, particularly if parts are required for any of the earlier models.

The frame number is located on the left hand side of the steering head and the engine number is stamped on the upper crankcase, immediately to the rear of the two left hand cylinders. Use only parts of genuine Honda manufacture.

Pattern parts are available, some of which originate from Japan, but in many instances they may have an adverse effect on performance and/or reliability. Furthermore the fitting of non-standard parts may invalidate the warranty. Honda do not operate a 'service exchange' scheme.

Some of the more expendable parts such as spark plugs, bulbs, tyres, oils and greases etc., can be obtained from accessory shops and motor factors, who have convenient opening hours, charge lower prices and can often be found not far from home. It is also possible to obtain parts on a Mail Order basis from a number of specialists who advertise regularly in the motor cycle magazines.

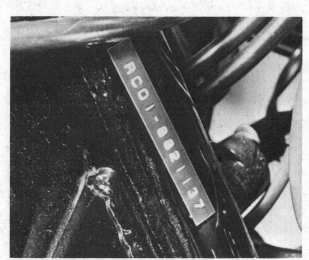

Location of frame number

Location of engine number

Safety first!

Professional motor mechanics are trained in safe working procedures. However enthusiastic you may be about getting on with the job in hand, do take the time to ensure that your safety is not put at risk. A moment's lack of attention can result in an accident, as can failure to observe certain elementary precautions.

There will always be new ways of having accidents, and the following points do not pretend to be a comprehensive list of all dangers; they are intended rather to make you aware of the risks and to encourage a safety-conscious approach to all work you carry out on your vehicle.

Essential DOs and DON'Ts

DON'T start the engine without first ascertaining that the transmission is in neutral.

DON'T suddenly remove the filler cap from a hot cooling system – cover it with a cloth and release the pressure gradually first, or you may get scalded by escaping coolant.

DON'T attempt to drain oil until you are sure it has cooled sufficiently to avoid scalding you.

DON'T grasp any part of the engine, exhaust or silencer without first ascertaining that it is sufficiently cool to avoid burning you.

DON'T allow brake fluid or antifreeze to contact the machine's paintwork or plastic components.

DON'T syphon toxic liquids such as fuel, brake fluid or antifreeze by mouth, or allow them to remain on your skin.

DON'T inhale dust – it may be injurious to health (see *Asbestos* heading).

DON'T allow any spilt oil or grease to remain on the floor – wipe it up straight away, before someone slips on it.

DON'T use ill-fitting spanners or other tools which may slip and cause injury.

DON'T attempt to lift a heavy component which may be beyond your capability – get assistance.

DON'T rush to finish a job, or take unverified short cuts.

DON'T allow children or animals in or around an unattended vehicle.

DON'T inflate a tyre to a pressure above the recommended maximum. Apart from overstressing the carcase and wheel rim, in extreme cases the tyre may blow off forcibly.

DO ensure that the machine is supported securely at all times. This is especially important when the machine is blocked up to aid wheel or fork removal.

DO take care when attempting to slacken a stubborn nut or bolt. It is generally better to pull on a spanner, rather than push, so that if slippage occurs you fall away from the machine rather than on to it.

DO wear eye protection when using power tools such as drill, sander, bench grinder etc.

DO use a barrier cream on your hands prior to undertaking dirty jobs – it will protect your skin from infection as well as making the dirt easier to remove afterwards; but make sure your hands aren't left slippery. Note that long-term contact with used engine oil can be a health hazard.

DO keep loose clothing (cuffs, tie etc) and long hair well out of the way of moving mechanical parts.

DO remove rings, wristwatch etc, before working on the vehicle – especially the electrical system.

DO keep your work area tidy – it is only too easy to fall over articles left lying around.

DO exercise caution when compressing springs for removal or installation. Ensure that the tension is applied and released in a controlled manner, using suitable tools which preclude the possibility of the spring escaping violently.

DO ensure that any lifting tackle used has a safe working load rating adequate for the job.

DO get someone to check periodically that all is well, when working alone on the vehicle.

DO carry out work in a logical sequence and check that everything is correctly assembled and tightened afterwards.

DO remember that your vehicle's safety affects that of yourself and others. If in doubt on any point, get specialist advice.

IF, in spite of following these precautions, you are unfortunate enough to injure yourself, seek medical attention as soon as possible.

Asbestos

Certain friction, insulating, sealing, and other products – such as brake linings, clutch linings, gaskets, etc – contain asbestos. *Extreme care must be taken to avoid inhalation of dust from such products since it is hazardous to health*. If in doubt, assume that they *do* contain asbestos.

Fire

Remember at all times that petrol (gasoline) is highly flammable. Never smoke, or have any kind of naked flame around, when working on the vehicle. But the risk does not end there – a spark caused by an electrical short-circuit, by two metal surfaces contacting each other, by careless use of tools, or even by static electricity built up in your body under certain conditions, can ignite petrol vapour, which in a confined space is highly explosive.

Always disconnect the battery earth (ground) terminal before working on any part of the fuel or electrical system, and never risk spilling fuel on to a hot engine or exhaust.

It is recommended that a fire extinguisher of a type suitable for fuel and electrical fires is kept handy in the garage or workplace at all times. Never try to extinguish a fuel or electrical fire with water.

Note: *Any reference to a 'torch' appearing in this manual should always be taken to mean a hand-held battery-operated electric lamp or flashlight. It does **not** mean a welding/gas torch or blowlamp.*

Fumes

Certain fumes are highly toxic and can quickly cause unconsciousness and even death if inhaled to any extent. Petrol (gasoline) vapour comes into this category, as do the vapours from certain solvents such as trichloroethylene. Any draining or pouring of such volatile fluids should be done in a well ventilated area.

When using cleaning fluids and solvents, read the instructions carefully. Never use materials from unmarked containers – they may give off poisonous vapours.

Never run the engine of a motor vehicle in an enclosed space such as a garage. Exhaust fumes contain carbon monoxide which is extremely poisonous; if you need to run the engine, always do so in the open air or at least have the rear of the vehicle outside the workplace.

The battery

Never cause a spark, or allow a naked light, near the vehicle's battery. It will normally be giving off a certain amount of hydrogen gas, which is highly explosive.

Always disconnect the battery earth (ground) terminal before working on the fuel or electrical systems.

If possible, loosen the filler plugs or cover when charging the battery from an external source. Do not charge at an excessive rate or the battery may burst.

Take care when topping up and when carrying the battery. The acid electrolyte, even when diluted, is very corrosive and should not be allowed to contact the eyes or skin.

If you ever need to prepare electrolyte yourself, always add the acid slowly to the water, and never the other way round. Protect against splashes by wearing rubber gloves and goggles.

Mains electricity and electrical equipment

When using an electric power tool, inspection light etc, always ensure that the appliance is correctly connected to its plug and that, where necessary, it is properly earthed (grounded). Do not use such appliances in damp conditions and, again, beware of creating a spark or applying excessive heat in the vicinity of fuel or fuel vapour. Also ensure that the appliances meet the relevant national safety standards.

Ignition HT voltage

A severe electric shock can result from touching certain parts of the ignition system, such as the HT leads, when the engine is running or being cranked, particularly if components are damp or the insulation is defective. Where an electronic ignition system is fitted, the HT voltage is much higher and could prove fatal.

Routine maintenance

Periodical routine maintenance is essential to keep the motorcycle in a peak and safe condition. Routine maintenance also saves money because it provides the opportunity to detect and remedy a fault before it develops further and causes more damage. Maintenance should be undertaken on either a calendar or mileage basis depending on whichever comes sooner. The period between maintenance tasks serves only as a guide since there are many variables eg; age of machine, riding technique and adverse conditions.

The maintenance instructions are generally those recommended by the manufacturer but are supplemented by additional tasks which, through practical experience, the author recommends should be carried out at the intervals suggested. The additional tasks are primarily of a preventative nature, which will assist in eliminating unexpected failure of a component or system, due to wear and tear, and increase safety margins when riding.

All the maintenance tasks are described in detail together with the procedures required for accomplishing them. If necessary, more general information on each topic can be found in the relevant Chapter within the main text.

Although no special tools are required for routine maintenance, a good selection of general workshop tools is essential. Included in the tools must be a range of metric ring or combination spanners, a selection of crosshead screwdrivers, and two pairs of circlip pliers, one external opening and the other internal opening. Additionally, owing to the extreme tightness of most casing screws on Japanese machines, an impact screwdriver, together with a choice of large or small cross-head screw bits, is absolutely indispensable. This is particularly so if the engine has not been dismantled since leaving the factory.

Weekly, or every 200 miles (320 km)

---

1 Tyres

Check the tyre pressures. Always check the pressure when the tyres are cold as the heat generated when the machine has been ridden can increase the pressures by as much as 8 psi, giving a totally inaccurate reading. Variations in pressure of as little as 2 psi may alter certain handling characteristics. It is therefore recommended that whatever type of pressure gauge is used, it should be checked occasionally to ensure accurate readings. Do not put absolute faith in 'free air' gauges at garages or petrol stations. They have been known to be in error.

Inspect the tyre treads for cracking or evidence that the outer rubber is leaving the inner cover. Also check the tyre walls for splitting or perishing. Carefully inspect the treads for stones, flints or shrapnel which may have become embedded and be slowly working their way towards the inner surface. Remove such objects with a suitable tool. The thing for getting stones out of horses hooves is ideal!

2 Battery

Check the electrolyte level in the battery and replenish, if necessary, with distilled water. Do not use tap water as this will reduce the life of the battery. If the battery is removed for filling, note the tracking of the battery breather pipe which should be replaced in the same position, ensuring that the pipe is not kinked or blocked. If the breather pipe is restricted and the battery overheats for any reason, the pressure produced may, in extreme cases, cause the battery case to fail and a liberal amount of sulphuric acid to be deposited on the electrical harness and frame parts.

3 Engine oil

Check the engine oil level by means of the dipstick incorporated in the filler plug which screws into the left-hand side of the crankcase. When taking the reading do not screw the plug into the casing; allow it to rest on the rim of the filler orifice. Replenish the engine oil with oil of the specified grade to the maximum level on the dipstick.

4 Electrical system

Check that the various bulbs are functioning properly, paying particular attention to the rear lamp. It is possible that one of the rear lamp or brake lamp filaments has failed but gone unnoticed. Check that the indicators and horn operate normally. Clean all lenses. If any of the fuses has blown recently, check that the source of the problem has been resolved and that the spare fuse has been renewed.

Correct engine oil level is between upper and lower marks on dipstick

5 Brake fluid

Check the hydraulic fluid level in the front brake master cylinder reservoir. Before removing the reservoir cap and diaphragm place the handlebars in such a position that the reservoir is approximately vertical. This will prevent spillage. The fluid should lie between the upper and lower lines on the reservoir body. Replenish, if necessary, with hydraulic brake fluid of the correct specification, which is DOT 3 (USA) or SAE-J1703. If the level of fluid in either of the reservoirs is excessively low, check the pads for wear. If the pads are not worn, suspect a fluid leakage in the system. This must be rectified immediately. In the case of machines fitted with a rear disc brake, check the fluid level as described above. The reservoir is located behind the right-hand side panel.

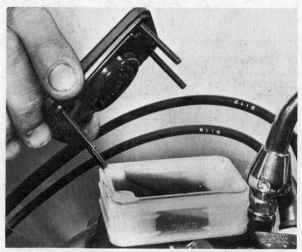

Top up brake fluid as necessary to maintain correct level

6 Safety inspection

Give the whole machine a close visual inspection, checking for loose nuts and fittings, frayed control cables and damaged brake hoses etc.

Monthly or every 600 miles (1000 km)

Complete all the checks listed under the previous maintenance interval heading and then carry out the following.

1 Final drive chain: lubrication and adjustment

The final drive chain is of the endless type, having no joining link in an effort to eliminate any tendency towards breakage. The rollers are equipped with an O-ring at each end which seals the lubricant inside and prevents the ingress of water or abrasive grit. It should not, however, be supposed that the need for lubrication is lessened. On the contrary, frequent but sparse lubrication is essential to minimise wear between the chain and sprockets. Honda recommend the use of SAE 80 or 90 gear oil for chain lubrication, but this will be of limited value due to the speed with which it is flung off. Conventional aerosol lubricants must be avoided because the propellant used will attack and damage the O-rings, but some of the newer types, such as PJL Blue Lable, are suitable for use on O-ring chains and are marked as such.

In particularly adverse weather conditions, or when touring, lubrication should be undertaken more frequently.

A final word of caution; the importance of chain lubrication cannot be overstressed in view of the cost of replacement, and the fact that a considerable amount of dismantling work, including swinging arm removal, will need to be undertaken should replacement be necessary.

Adjust the chain after lubrication, so that there is approximately 15-25 mm slack in the middle of the lower run. Always check with the chain at the tightest point as a chain rarely wears evenly during service.

Adjustment is accomplished after placing the machine on the centre stand and slackening the wheel nut, so that the wheel can be drawn backwards by means of the drawbolt adjusters in the fork ends.

The torque arm nuts and the rear brake adjuster (drum braked models) must also be slackened during this operation. Adjust the drawbolts an equal amount to preserve wheel alignment. The forks ends are clearly marked with a series of parallel lines above the adjusters, to provide a simple visual check.

2 Brake wear

Check that when applied, the rear brake wear indicator is within the usable range scale marked on the brake plate. The front disc brake pads should also be examined for wear, and to this end are marked with a red line denoting the maximum wear limit. If necessary, change the pads and/or brake shoes, referring to Chapter 5 for details. Look also for signs of staining on the friction material. This may be caused by leakage from the fork leg or from the caliper seals; in either case attention must be given to locating and rectifying the source of the leak.

3 Wheel condition – wire spoked types

Check the spoke tension by gently tapping each one with a metal object. A loose spoke is identifiable by the low pitch noise generated. If any spoke needs considerable tightening, it will be necessary to remove the tyre and inner tube in order to file down the protruding spoke end. This will prevent the spoke from chafing through the rim band and piercing the inner tube. Rotate the wheel and test for rim runout. Excessive runout will cause handling problems and should be corrected by tightening or loosening the relevant spokes. Care must be taken, since altering the tension in the wrong spokes may create more problems.

4 Further maintenance checks

The following areas should be given a cursory check, taking remedial action where required. Check the electrical system, plus the headlamp beam alignment. Check the various nuts, bolts and screws for security, tightening where necessary. Check the front and rear suspension for smooth operation. Check the steering head bearings for free play. Examine all control cables and hydraulic lines, renewing any which appear worn or frayed.

---

6 monthly or every 3600 miles (6000 km)

Complete all the checks in the preceding maintenance schedules and then carry out the following.

1 Engine oil and filter renewal

Run the engine until normal operating temperature is reached to ensure that the old oil drains quickly and completely. Place the machine on its centre stand and position a drain tray or bowl of about 1 gallon capacity beneath the sump drain plug. Release the drain bolt and filler plug, and allow the engine oil to drain.

Slacken the oil filter bolt and remove the filter housing and element. Note that some residual oil will be released, and some provision must be made to catch this. When all the oil has drained, clean the area around the drain plug and filter housing, and the inside of the filter bowl. Refit the drain plug and filter assembly. Fill the crankcase with 3.5 litre (6.0 Imp pint, 3.7 US quart) of the recommended engine oil, then run the engine for two or three minutes, checking for signs of leakage around the drain plug and filter. Stop the engine and check the oil level, adding oil where necessary to bring the level to maximum on the dipstick.

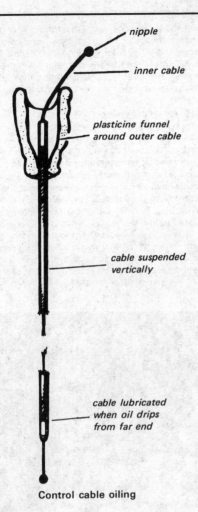

nipple

inner cable

plasticine funnel
around outer cable

cable suspended
vertically

cable lubricated
when oil drips
from far end

Control cable oiling

Oil filter housing and element

Replenish engine with oil until the correct amount is shown on dipstick

2 Valve clearances

It is important that valve clearances be maintained otherwise damage, or at best poor performance and noisy operation, will occur. To gain access to the camshafts, it will be necessary to detach the fuel tank and the H-shaped camshaft cover to expose the two camshafts and their associated components. Note that the engine should be **cold** during the clearance check.

Each valve is operated by a bucket-shaped follower which contains a shim to provide the correct clearance between it and the cam lobe. The gap should be measured with the peak of the cam lobe uppermost, at which point it should be possible to insert a feeler gauge between the bucket top and the cam lobe. The specified clearance is,

$$0.08 \text{ mm} \begin{array}{l} + 0.05 \\ - 0.02 \end{array} \quad (0.003 \text{ in} \begin{array}{l} + 0.002 \\ - 0.001 \end{array})$$

for both the inlet and the exhaust valves. This gives a permissible range of 0.06 – 0.13 mm (0.002 – 0.005 in) in each case.

It will be necessary to set the camshafts in the correct position, as described below. Working from the right-hand side of the machine, turn the crankshaft slowly in a clockwise direction until the index mark on the exhaust camshaft end aligns with the front section of the gasket face. Check and note the clearance of the exhaust valves of cylinders 1 and 3.

Rotate the camshafts through 90°, by turning the crankshaft clockwise through 180°, then repeat the check on the inlet valves of cylinders 1 and 3.

Turn the camshafts through another 90° and check the exhaust valves of cylinders 2 and 4. Finally, rotate the camshafts by a further 90°, and check the clearance of the inlet valves of cylinders 2 and 4.

Check valve clearances using a feeler gauge

Check the list of clearances against the specified clearance limits to see which, if any, require attention. Adjustment shims are available in 0.05 mm increments, from 2.30 mm to 3.50 mm. Thus if a clearance of 0.15 mm is found and the existing shim is 2.45 mm thick, it will be necessary to fit a 2.50 mm shim to bring the clearance within limits, to 0.10 mm.

To change the shims it will be necessary to keep the appropriate pair of cam followers depressed so that the shim(s) can be withdrawn from the recess in the cam follower top(s). It is strongly recommended that the Honda tool, No 07964–4220001 is purchased for this purpose, because it allows the job to be accomplished quickly and at no risk to the machine or operator. Turn the crankshaft until the valve in question is fully open, and insert the tool between the camshaft and the appropriate pair of valves. Turn the crankshaft through 360° to position the camshaft lobe clear of the valves.

Note that care must be taken not to rotate the crankshaft so far that the opposing pair of valves is opened. If this happens, the inlet and exhaust valve heads could meet, causing damage to both.

With the appropriate valve held open, the adjustment shim can be dislodged with a small screwdriver and lifted clear using tweezers or pointed-nose pliers. All but the No 2 cylinder exhaust valve shims can be removed from the sparking plug side of the camshaft. The latter must be removed from the front of the cylinder head.

The offending shim can now be measured with a micrometer, and the appropriate replacement fitted, having referred to the list of clearances made earlier. Do not forget to recheck the setting after the holding tool has been removed.

In the event that a new shim will not give the required clearance, it is likely that the valve seat and/or valve is in need of renewal. On no account attempt to grind down existing shims or pack them with sheet shim material in an attempt to save the cost of new shims. The risk of failure in service, and the consequent damage to the engine, makes this a very false economy. For details of shim sizes refer to the accompanying table of sizes.

3 Cam chain tensioner adjustment

The camshaft drive and connecting chains are tensioned by sprung blade assemblies mounted in the chain tunnel and across the cylinder head. The tensioners are semi-automatic, adjusting to the correct tension when the lock nuts are released.

It may be noted that the procedure described here differs slightly from that given in the owners handbook. This is because a revised procedure has been found to give better results, and is now recommended by Honda for all models.

Camshaft drive chain

Start the engine and allow it to idle. Slacken both of the small domed nuts at the rear of the cylinder block to allow the tensioner to assume the correct position. Tighten the two nuts to lock the adjustment.

Camshaft connecting chain

With the engine idling, slacken the locknut and bolt at the front of the cylinder block by $\frac{1}{2}$ turn, to allow the horizontal tensioner to assume the correct position. Tighten the bolt carefully to hold the adjustment, taking care not to over-tighten it. Retighten the locknut.

If either chain remains noisy it is likely that it has stretched to the point where it requires renewal. Refer to Chapter 1 for details of the renewal procedure.

4 Sparking plugs

Remove, clean and adjust the sparking plugs. Carbon and other deposits can be removed, using a wire brush, and emery paper or a file used to clean the electrodes prior to adjusting the gaps. Probably the best method of sparking plug cleaning is by having them shot blasted in a special machine. This type of machine is used by most garages. If the outer electrode of a plug is excessively worn (indicated by a step in the underside)

the plug should be renewed. Adjust the points gap on each plug by bending the outer electrode only, so that the gap is within the range 0.6 – 0.7 mm (0.024 – 0.028 in). Before replacing the plugs, smear the threads with graphited grease; this will aid subsequent removal. If replacement plugs are required the correct types are listed at the end of this Section.

5 Air filter cleaning

Access to the filter is gained after removing the left-hand side panel. Remove the air cleaner casing cover by releasing the two screws which retain it. The element is secured by a leaf spring and can be removed after the latter has been pulled clear.

Tap the element gently to remove any loose dust and then use an air hose to remove the remainder of the dust. Apply the air current from the inside of the element only. If an air hose is not available, a tyre pump can be utilised instead. If the corrugated paper element is damp, oily or beginning to disintegrate, it must be renewed. Do not run the engine with the element removed as the weak mixture caused may result in engine overheating and damage to the cylinders and pistons. A weak mixture can also result if the rubber sealing rings on the element are perished or omitted.

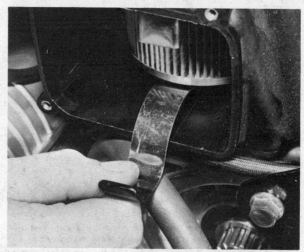

Air filter can be removed by pulling the leaf spring clear ...

... and withdrawing the element

VALVE SHIM SELECTION CHART

STANDARD VALVE CLEARANCE = 0.08 $^{+0.05}_{-0.02}$ mm

PRESENT SHIM SIZE mm

VALVE CLEARANCE mm	2.30	2.35	2.40	2.45	2.50	2.55	2.60	2.65	2.70	2.75	2.80	2.85	2.90	2.95	3.00	3.05	3.10	3.15	3.20	3.25	3.30	3.35	3.40	3.45	3.50
0.01–0.05		2.30	2.35	2.40	2.45	2.50	2.55	2.60	2.65	2.70	2.75	2.80	2.85	2.90	2.95	3.00	3.05	3.10	3.15	3.20	3.25	3.30	3.35	3.40	3.45
0.06–0.13	*NO CHANGE REQUIRED (SPECIFIED CLEARANCE)*																								
0.14–0.16	2.35	2.40	2.45	2.50	2.55	2.60	2.65	2.70	2.75	2.80	2.85	2.90	2.95	3.00	3.05	3.10	3.15	3.20	3.25	3.30	3.35	3.40	3.45	3.50	
0.17–0.21	2.40	2.45	2.50	2.55	2.60	2.65	2.70	2.75	2.80	2.85	2.90	2.95	3.00	3.05	3.10	3.15	3.20	3.25	3.30	3.35	3.40	3.45	3.50		
0.22–0.26	2.45	2.50	2.55	2.60	2.65	2.70	2.75	2.80	2.85	2.90	2.95	3.00	3.05	3.10	3.15	3.20	3.25	3.30	3.35	3.40	3.45	3.50			
0.27–0.31	2.50	2.55	2.60	2.65	2.70	2.75	2.80	2.85	2.90	2.95	3.00	3.05	3.10	3.15	3.20	3.25	3.30	3.35	3.40	3.45	3.50				
0.32–0.36	2.55	2.60	2.65	2.70	2.75	2.80	2.85	2.90	2.95	3.00	3.05	3.10	3.15	3.20	3.25	3.30	3.35	3.40	3.45	3.50					
0.37–0.41	2.60	2.65	2.70	2.75	2.80	2.85	2.90	2.95	3.00	3.05	3.10	3.15	3.20	3.25	3.30	3.35	3.40	3.45	3.50						
0.42–0.46	2.65	2.70	2.75	2.80	2.85	2.90	2.95	3.00	3.05	3.10	3.15	3.20	3.25	3.30	3.35	3.40	3.45	3.50							
0.47–0.51	2.70	2.75	2.80	2.85	2.90	2.95	3.00	3.05	3.10	3.15	3.20	3.25	3.30	3.35	3.40	3.45	3.50								
0.52–0.56	2.75	2.80	2.85	2.90	2.95	3.00	3.05	3.10	3.15	3.20	3.25	3.30	3.35	3.40	3.45	3.50									
0.57–0.61	2.80	2.85	2.90	2.95	3.00	3.05	3.10	3.15	3.20	3.25	3.30	3.35	3.40	3.45	3.50										
0.62–0.66	2.85	2.90	2.95	3.00	3.05	3.10	3.15	3.20	3.25	3.30	3.35	3.40	3.45	3.50											
0.67–0.71	2.90	2.95	3.00	3.05	3.10	3.15	3.20	3.25	3.30	3.35	3.40	3.45	3.50												
0.72–0.76	2.95	3.00	3.05	3.10	3.15	3.20	3.25	3.30	3.35	3.40	3.45	3.50													
0.77–0.81	3.00	3.05	3.10	3.15	3.20	3.25	3.30	3.35	3.40	3.45	3.50														
0.82–0.86	3.05	3.10	3.15	3.20	3.25	3.30	3.35	3.40	3.45	3.50															
0.87–0.91	3.10	3.15	3.20	3.25	3.30	3.35	3.40	3.45	3.50																
0.92–0.96	3.15	3.20	3.25	3.30	3.35	3.40	3.45	3.50																	
0.97–1.01	3.20	3.25	3.30	3.35	3.40	3.45	3.50																		
1.02–1.06	3.25	3.30	3.35	3.40	3.45	3.50																			
1.07–1.11	3.30	3.35	3.40	3.45	3.50																				
1.12–1.16	3.35	3.40	3.45	3.50																					
1.17–1.21	3.40	3.45	3.50																						
1.22–1.26	3.45	3.50																							
1.27–1.31	3.50																								

EX → VALVE CLEARANCE 0.14–0.16 / EX → PRESENT SHIM 2.50 → 2.55

NOTE

(1) Measure the valve clearance while the engine is cold.

(2) For shim replacement, see text.

(3) Measure old and new shims with a micrometer.

(4) The chart is for reference purpose only. After installing new shims, recheck the valve clearance and adjust if necessary. Before rechecking, rotate the camshafts several times to seat the shims in the lifters.

(5) If the shim thickness required exceeds 3.5 mm, there is carbon build-up on the valve seat. Remove the carbon and reface the seat.

EXAMPLE:

1. Measure valve clearance = 0.16 mm
2. Measure present shim size = 2.50 mm
3. Refer to chart. (See shaded columns)
4. Replacement shim size = 2.55 mm

Valve clearance shim selection chart

6   Carburettor adjustment

The following points should be checked, and if necessary, adjusted. Note that adjustments should not be made by way of experimentation — if all is well, leave the carburettors alone. In practice it will be found that carburation adjustments are maintained with reasonable accuracy over quite long periods.

Run the engine to raise the temperature to normal, preferably by riding it for 10 – 15 minutes. Place the machine on its centre stand and allow the engine to idle. Check the idle speed, which should be 1000 ± 100 rpm. If necessary, move the throttle stop screw to bring the idle speed within limits.

Carburettor synchronisation requires the use of a vacuum gauge set. If this is not available, **do not** attempt adjustment, but take the machine to a Honda dealer to have this operation carried out.

If the vacuum gauge set is available, proceed as follows. Remove the dualseat and petrol tank so that access can be gained to the carburettors. Using a suitable length of feed pipe, reconnect the petrol tank with the carburettors, so that the petrol flow can be maintained. The petrol tank must be placed above the level of the carburettors. Connect the vacuum gauges to the engine.

Start the engine and allow it to run until normal working temperature has been reached. This should take 10 – 15 minutes. Set the throttle so that an engine speed of 1000 ± 100 rpm is maintained. If the readings on the vacuum gauges vary by more than 60 mm Hg (2.4 in Hg) it will be necessary to adjust the synchronising screws to bring the carburettors within limits. Note that if the readings on the gauges fluctuate wildly, it is likely that the gauges require heavier damping. Refer to the gauge manufacturer's instructions on setting up procedures.

The No 2 carburettor (second from left) is regarded as the base instrument; that is, it is non-adjustable and the remaining three carburettors must be adjusted to it. Honda produce a special combined screwdriver and socket spanner for dealing with the synchronising screws (Part number 07908–4220100). Its use makes the procedure easier, but it is not essential. Slacken the locknut of the adjuster concerned, then turn the latter, noting the effect on the gauge reading. When the reading is as close as possible to that of the No 2 carburettor, hold the adjuster screw and retighten the locknut. Repeat the procedure on the remaining carburettors.

To check the operation of the throttle twistgrip and cables, turn the fuel supply off and allow the engine to idle until it stalls due to the lack of fuel. This will prevent subsequent flooding problems where accelerator pumps are fitted.

The throttle twistgrip should have about 2 – 6 mm free play measured at the inner, flanged, edge. Coarse adjustment can be made by moving the lower adjuster by the required amount. Further fine adjustment is made by means of the upper adjuster.

7   Side stand: pad renewal

The side stand is fitted with a rubber pad which will gradually wear down with use. Check the pad condition and renew it when it nears the raised wear line. The old pad can be released by removing the single retaining bolt and the new item fitted by reversing the dismantling sequence.

Yearly or every 7200 miles (12000 km)

Carry out the operations listed in the previous Sections, then carry out the following.

1   Renew the sparking plugs

2   Renew the air filter element

In addition, the machine should be given a close visual examination, checking the numerous details not covered by the normal maintenance schedule. Check for signs of corrosion or rusting around the frame and cycle parts, taking the appropriate remedial action where required.

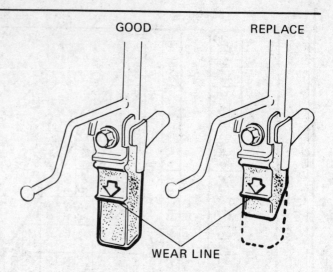

Checking side stand pad wear

Additional routine maintenance

---

1   Brake pads: examination and replacement

The rate of brake pad wear is dependent on the conditions under which the machine operates, weight carried and the style of riding, consequently it is difficult to advise on specific inspection intervals. Whatever inspection interval is chosen bear in mind that the rate of wear will not be constant.

To check wear on the front brake pads examine the pads through the small window in the main caliper units. If the red mark on the periphery of any pad has been reached, both pads in that set must be renewed. The rate of wear of the two sets are similar so it is probable that they will require renewal at the same time in any case.

Check the rear brake pad wear after removing the plastic caliper cover from position. If the red tongues on the pads have closed together sufficiently that they are within the area marked red on the caliper, they must be renewed.

2   Brake pad renewal

Remove each brake set individually, using an identical procedure as follows:

Unscrew the two bolts that pass into the caliper body and secure the body to the support bracket. Lift the caliper body off the support bracket, still interconnected with the hydraulic hose.

Lift the old pads out. Install the new pads and also the shim which fits against the outer face of the outer pad. The shim must be fitted so that the arrow is in the forward-most position, pointing in an upward direction. Refit the caliper halves and replace the socket screws. It may be necessary to push the caliper cylinder piston inwards to give the necessary clearance. If required, the bleed screw on the caliper can be slackened at the same time as the piston is pushed inwards. This will allow a small amount of fluid to seep out and the piston to move. Place a rag around the bleed screw to prevent the fluid leaking onto the caliper unit. Operate the brake lever, after pad replacement, to check free movement of the pads and to allow the pads to self-adjust.

3   Rear brake adjustment – drum brake models

Adjustment of the rear brake should be carried out when required, the intervals being dependent on riding style and usage. The adjustment is carried out by turning the nut on the rear end of the brake operating rod. Precise adjustment is a matter of personal choice but it should be ensured that the brake does not bind when the foot pedal is in the fully returned position. A brake wear indicator is fitted to the brake actuating arm on the back plate. If, when the brake is applied fully, the arrow on the arm is in line with the cast-in index mark on the back plate, the brake linings are worn sufficiently to require renewal.

4 Clutch adjustment

In common with brake pad wear, clutch wear and the resultant necessary adjustment depends on operating conditions and the style of riding. Adjust the clutch, when necessary, as follows.

Check the clutch free play at the ball-end of the handlebar lever. The lever should move 10 – 20 mm ($\frac{3}{8}$ – $\frac{3}{4}$ in) before the clutch begins to lift. If the free play is incorrect, the cable may be adjusted by means of the adjuster screws at both ends of the cable. The lower adjuster is used for coarse adjustments and the upper adjuster for finer running adjustments. If the upper adjuster threads project more than 8 mm (0.3 in) from the lever stock, there is some danger of the adjuster breaking out of the stock boss. To prevent this, screw the adjuster in fully and then make the adjustment for free play only using the lower adjuster to take up the excess slack.

Clutch adjustment mechanism for coarse adjustments

Quick glance
maintenance adjustments and capacities

Engine/transmission oil capacity
  Dry .... 4.5 litre (7.92 Imp pint/4.7 US quart)
  At oil change .... 3.5 litre (6.00 Imp pint/3.7 US quart)

Sparking plug gap .... 0.6 – 0.7 mm (0.024 – 0.028 in)

Front fork oil capacity
  Dry .... 175 cc
  At oil change .... 155 cc

Tyre pressures
  Front:
    CB750 K, LTD and F .... 28 psi (2.0 kg/cm²)
    CB900 F .... 32 psi (2.25 kg/cm²)
  Rear (solo):
    CB750 pre-1980 and F .... 32 psi (2.25 kg/cm²)
    CB750 K 1980 and LTD .... 28 psi (2.0 kg/cm²)
    CB900 .... 36 psi (2.50 kg/cm²)
  Rear (with more than 90 kg [200 lb] load):
    All models .... 40 psi (2.80 kg/cm²)

Recommended lubricants

Components	Lubricant
Engine/transmission	
General, all-temperature use	SAE 10W/40
Above 15°C (60°F)	SAE 30
−10° to + 15°C (15° – 60°F)	SAE 20 or 20W
Above −10°C (15°F)	SAE 20W/50
Below 0°C (32°F)	SAE 10W
Front forks	Automatic transmission fluid (ATF)
Chain	SAE 80 or 90 gear oil
General lubrication	Light machine oil
Wheel bearings	High melting point grease
Swinging arm	High melting point grease
Component lubrication during engine reassembly (see text)	Molybdenum disulphide grease

Tools and working facilities

The first priority when undertaking maintenance or repair work of any sort on a motorcycle is to have a clean, dry, well-lit working area. Work carried out in peace and quiet in the well-ordered atmosphere of a good workshop will give more satisfaction and much better results than can usually be achieved in poor working conditions. A good workshop must have a clean flat workbench or a solidly constructed table of convenient working height. The workbench or table should be equipped with a vice which has a jaw opening of at least 4 in (100 mm). A set of jaw covers should be made from soft metal such as aluminium alloy or copper, or from wood. These covers will minimise the marking or damaging of soft or delicate components which may be clamped in the vice. Some clean, dry, storage space will be required for tools, lubricants and dismantled components. It will be necessary during a major overhaul to lay out engine/gearbox components for examination and to keep them where they will remain undisturbed for as long as is necessary. To this end it is recommended that a supply of metal or plastic containers of suitable size is collected. A supply of clean, lint-free, rags for cleaning purposes and some newspapers, other rags, or paper towels for mopping up spillages should also be kept. If working on a hard concrete floor note that both the floor and one's knees can be protected from oil spillages and wear by cutting open a large cardboard box and spreading it flat on the floor under the machine or workbench. This also helps to provide some warmth in winter and to prevent the loss of nuts, washers, and other tiny components which have a tendency to disappear when dropped on anything other than a perfectly clean, flat, surface.

Unfortunately, such working conditions are not always available to the home mechanic. When working in poor conditions it is essential to take extra time and care to ensure that the components being worked on are kept scrupulously clean and to ensure that no components or tools are lost or damaged.

A selection of good tools is a fundamental requirement for anyone contemplating the maintenance and repair of a motor vehicle. For the owner who does not possess any, their purchase will prove a considerable expense, offsetting some of the savings made by doing-it-yourself. However, provided that the tools purchased meet the relevant national safety standards and are of good quality, they will last for many years and prove an extremely worthwhile investment.

To help the average owner to decide which tools are needed to carry out the various tasks detailed in this manual, we have compiled three lists of tools under the following headings: *Maintenance and minor repair*, *Repair and overhaul*, and *Specialized*. The newcomer to practical mechanics should start off with with the simpler jobs around the vehicle. Then, as his confidence and experience grow, he can undertake more difficult tasks, buying extra tools as and when they are needed. In this way, a *Maintenance and minor repair* tool kit can be built-up into a *Repair and overhaul* tool kit over a considerable period of time without any major cash outlays. The experienced home mechanic will have a tool kit good enough for most repair and overhaul procedures and will add tools from the specialized category when he feels the expense is justified by the amount of use these tools will be put to.

It is obviously not possible to cover the subject of tools fully here. For those who wish to learn more about tools and their use there is a book entitled *Motorcycle Workshop Practice Manual* (Book no 1454) available from the publishers of this manual.

As a general rule, it is better to buy the more expensive, good quality tools. Given reasonable use, such tools will last for a very long time, whereas the cheaper, poor quality, item will wear out faster and need to be renewed more often, thus nullifying the original saving. There is also the risk of a poor quality tool breaking while in use, causing personal injury or expensive damage to the component being worked on.

For practically all tools, a tool factor is the best source since he will have a very comprehensive range compared with the average garage or accessory shop. Having said that, accessory shops often offer excellent quality tools at discount prices, so it pays to shop around. There are plenty of tools around at reasonable prices, but always aim to purchase items which meet the relevant national safety standards. If in doubt, seek the advice of the shop proprietor or manager before making a purchase.

The basis of any toolkit is a set of spanners. While open-ended spanners with their slim jaws, are useful for working on awkwardly-positioned nuts, ring spanners have advantages in that they grip the nut far more positively. There is less risk of the spanner slipping off the nut and damaging it, for this reason alone ring spanners are to be preferred. Ideally, the home mechanic should acquire a set of each, but if expense rules this out a set of combination spanners (open-ended at one end and with a ring of the same size at the other) will provide a good compromise. Another item which is so useful it should be considered an essential requirement for any home mechanic is a set of socket spanners. These are available in a variety of drive sizes. It is recommended that the $\frac{1}{2}$-inch drive type is purchased to begin with as although bulkier and more expensive than the $\frac{3}{8}$-inch type, the larger size is far more common and will accept a greater variety of torque wrenches, extension pieces and socket sizes. The socket set should comprise sockets of sizes between 8 and 24 mm, a reversible ratchet drive, an extension bar of about 10 inches in length, a spark plug socket with a rubber insert, and a universal joint. Other attachments can be added to the set at a later date.

Maintenance and minor repair tool kit

Set of spanners 8 – 24 mm
Set of sockets and attachments
Spark plug spanner with rubber insert – 10, 12, or 14 mm as appropriate
Adjustable spanner
C-spanner/pin spanner
Torque wrench (same size drive as sockets)
Set of screwdrivers (flat blade)
Set of screwdrivers (cross-head)
Set of Allen keys 4 – 10 mm
Impact screwdriver and bits
Ball pein hammer – 2 lb
Hacksaw (junior)
Self-locking pliers – Mole grips or vice grips
Pliers – combination
Pliers – needle nose
Wire brush (small)
Soft-bristled brush
Tyre pump
Tyre pressure gauge
Tyre tread depth gauge
Oil can
Fine emery cloth
Funnel (medium size)
Drip tray
Grease gun
Set of feeler gauges
Brake bleeding kit
Strobe timing light
Continuity tester (dry battery and bulb)
Soldering iron and solder
Wire stripper or craft knife
PVC insulating tape
Assortment of split pins, nuts, bolts, and washers

Repair and overhaul toolkit

The tools in this list are virtually essential for anyone undertaking major repairs to a motorcycle and are additional to the tools listed above. Concerning Torx driver bits, Torx screws are encountered on some of the more modern machines where their use is restricted to fastening certain components inside the engine/gearbox unit. It is therefore recommended that if Torx bits cannot be borrowed from a local dealer, they are purchased individually as the need arises. They are not in regular use in the motor trade and will therefore only be available in specialist tool shops.

> *Plastic or rubber soft-faced mallet*
> *Torx driver bits*
> *Pliers – electrician's side cutters*
> *Circlip pliers – internal (straight or right-angled tips are available)*
> *Circlip pliers – external*
> *Cold chisel*
> *Centre punch*
> *Pin punch*
> *Scriber*
> *Scraper (made from soft metal such as aluminium or copper)*
> *Soft metal drift*
> *Steel rule/straight edge*
> *Assortment of files*
> *Electric drill and bits*
> *Wire brush (large)*
> *Soft wire brush (similar to those used for cleaning suede shoes)*
> *Sheet of plate glass*
> *Hacksaw (large)*
> *Valve grinding tool*
> *Valve grinding compound (coarse and fine)*
> *Stud extractor set (E-Z out)*

Specialized tools

This is not a list of the tools made by the machine's manufacturer to carry out a specific task on a limited range of models. Occasional references are made to such tools in the text of this manual and, in general, an alternative method of carrying out the task without the manufacturer's tool is given where possible. The tools mentioned in this list are those which are not used regularly and are expensive to buy in view of their infrequent use. Where this is the case it may be possible to hire or borrow the tools against a deposit from a local dealer or tool hire shop. An alternative is for a group of friends or a motorcycle club to join in the purchase.

> *Valve spring compressor*
> *Piston ring compressor*
> *Universal bearing puller*

> *Cylinder bore honing attachment (for electric drill)*
> *Micrometer set*
> *Vernier calipers*
> *Dial gauge set*
> *Cylinder compression gauge*
> *Vacuum gauge set*
> *Multimeter*
> *Dwell meter/tachometer*

Care and maintenance of tools

Whatever the quality of the tools purchased, they will last much longer if cared for. This means in practice ensuring that a tool is used for its intended purpose; for example screwdrivers should not be used as a substitute for a centre punch, or as chisels. Always remove dirt or grease and any metal particles but remember that a light film of oil will prevent rusting if the tools are infrequently used. The common tools can be kept together in a large box or tray but the more delicate, and more expensive, items should be stored separately where they cannot be damaged. When a tool is damaged or worn out, be sure to renew it immediately. It is false economy to continue to use a worn spanner or screwdriver which may slip and cause expensive damage to the component being worked on.

Fastening systems

Fasteners, basically, are nuts, bolts and screws used to hold two or more parts together. There are a few things to keep in mind when working with fasteners. Almost all of them use a locking device of some type; either a lock washer, lock nut, locking tab or thread adhesive. All threaded fasteners should be clean, straight, have undamaged threads and undamaged corners where the hexagon head fits. Develop the habit of replacing all damaged nuts and bolts with new ones.

Rusted nuts and bolts should be treated with a rust penetrating fluid to ease removal and prevent breakage. After applying the rust penetrant, let it 'work' for a few minutes before trying to loosen the nut or bolt. Badly rusted fasteners may have to be chiseled off or removed with a special nut breaker, available at tool shops.

Flat washers and lock washers, when removed from an assembly should always be replaced exactly as removed. Replace any damaged washers with new ones. Always use a flat washer between a lock washer and any soft metal surface (such as aluminium), thin sheet metal or plastic. Special lock nuts can only be used once or twice before they lose their locking ability and must be renewed.

If a bolt or stud breaks off in an assembly, it can be drilled out and removed with a special tool called an E-Z out. Most dealer service departments and motorcycle repair shops can perform this task, as well as others (such as the repair of threaded holes that have been stripped out).

Standard torque settings

Specific torque settings will be found at the end of the specifications section of each chapter. Where no figure is given, bolts should be secured according to the table below.

Fastener type (thread diameter)	kgf m	lbf ft
5 mm bolt or nut	0.45 – 0.6	3.5 – 4.5
6 mm bolt or nut	0.8 – 1.2	6 – 9
8 mm bolt or nut	1.8 – 2.5	13 – 18
10 mm bolt or nut	3.0 – 4.0	22 – 29
12 mm bolt or nut	5.0 – 6.0	36 – 43
5 mm screw	0.35 – 0.5	2.5 – 3.6
6 mm screw	0.7 – 1.1	5 – 8
6 mm flange bolt	1.0 – 1.4	7 – 10
8 mm flange bolt	2.4 – 3.0	17 – 22
10 mm flange bolt	3.0 – 4.0	22 – 29

Chapter 1 Engine, clutch and gearbox

For modifications, and information relating to later models, see Chapter 7

Contents

Specifications

Note: *Specifications are given for the 900 model only where they differ from the 750 model*

Model	CB750	CB900F
Engine		
Type	Four-cylinder air-cooled dohc	Four-stroke
Bore	62 mm (2.44 in)	64.5 mm (2.54 in)
Stroke	62 mm (2.44 in)	69.0 mm (2.71 in)
Capacity	749 cc (45.67 cu in)	901 cc (54.9 cu in)
Compression ratio	9.0:1	8.8:1
Valve clearances (engine cold)		
Inlet and exhaust	0.06 – 0.13 mm (0.002 – 0.005 in)	

Valve timing

Inlet opens at	5° BTDC at 1 mm lift, 58° BTDC at 0 lift	10° BTDC at 1 mm lift, 62° BTDC at 0 lift
Inlet closes at	35° ABDC at 1 mm lift, 101° ABDC at 0 lift	35° ABDC at 1 mm lift, 90° ABDC at 0 lift
Exhaust opens at	35° BBDC at 1 mm lift, 87° BBDC at 0 lift	40° BBDC at 1 mm lift, 93° BBDC at 0 lift
Exhaust closes at	5° ATDC at 1 mm lift, 72° ATDC at 0 lift	5° ATDC at 1 mm lift, 70° ATDC at 0 lift

Valve and springs

Seat angle	45°
Seat width	0.99 – 1.27 mm (0.039 – 0.050 in)
Service limit	1.5 mm (0.06 in)
Valve stem OD:	
Inlet valve	5.475 – 5.490 (0.2156 – 0.2161 in)
Service limit	5.470 mm (0.2150 in)
Exhaust valve	5.445 – 5.470 mm (0.2148 – 0.2154 in)
Service limit	5.440 mm (0.2140 in)
Valve guide ID:	
Inlet valve	5.500 – 5.515 mm (0.2165 – 0.2171 in)
Service limit	5.540 mm (0.2180 in)
Exhaust valve	5.500 – 5.515 mm (0.2165 – 0.2171 in)
Service limit	5.540 mm (0.2180 in)
Stem to guide clearance service limit:	
Inlet valve	0.07 mm (0.003 in)
Exhaust valve	0.09 mm (0.004 in)
Spring free length:	
Inlet (outer)	43.9 mm (1.73 in)
Service limit	42.5 mm (1.67 in)
Inlet (inner)	40.7 mm (1.60 in)
Service limit	39.8 mm (1.57 in)
Exhaust (outer)	43.9 mm (1.73 in)
Service limit	42.5 mm (1.67 in)
Exhaust (inner)	40.7 mm (1.60 in)
Service limit	39.8 mm (1.57 in)
Spring pressure at specified lengths:	
Inlet (outer)	12.6 – 14.6 kg @ 37.5 mm (27.78 – 32.19 lb @ 1.48 in)
Service limit	12 kg (26.46 lb)
Inlet (inner)	6.39 – 7.81 kg @ 34.5 mm (14.087 – 17.218 lb @ 1.36 in)
Service limit	6.0 kg (13.23 lb)
Exhaust (outer)	12.6 – 14.6 kg @ 37.5 mm (27.78 – 32.19 lb @ 1.48 in)
Service limit	12 kg (26.46 lb)
Exhaust (inner)	6.39 – 7.81 kg @ 34.5 mm (14.087 – 17.218 lb @ 1.36 in)
Service limit	6.0 kg (13.23 lb)

Cam followers and camshafts

	750 models	900 model
Cam follower OD	27.972 – 27.993 mm (1.1013 – 1.1021 in)	
Service limit	27.96 mm (1.101 in)	
Cam follower bore ID	28.000 – 28.016 mm (1.1024 – 1.1030 in)	
Service limit	28.04 mm (1.104 in)	
Cam follower to bore maximum clearance	0.07 mm (0.003 in)	
Camshaft overall height:		
Inlet	37.000 – 37.160 mm (1.4567 – 1.4630 in)	37.420 – 37.580 mm (1.4732 – 1.4795 in)
Service limit	36.9 mm (1.45 in)	37.3 mm (1.47 in)
Exhaust	37.500 – 37.660 mm (1.4763 – 1.4827 in)	37.920 – 38.080 mm (1.4929 – 1.4992 in)
Service limit	37.4 mm (1.47 in)	37.8 mm (1.49 in)
Camshaft to bearing cap clearance:		
Cap A, E, F and L	0.040 – 0.082 mm (0.0016 – 0.0032 in)	
Service limit	0.13 mm (0.0051 in)	
Cap B, C, H and J	0.085 – 0.139 (0.0033 – 0.0055 in)	
Service limit	0.19 mm (0.0075 in)	
All remaining caps	0.062 – 0.109 mm (0.0024 – 0.0043 in)	
Service limit	0.16 mm (0.0063 in)	
Maximum camshaft run-out	0.05 mm (0.002 in)	

Camshaft chains
Camshaft connecting chain length (see text) .... 175.70 – 175.92 mm (6.917 – 6.926 in)
Service limit .... 177.1 mm (6.97 in)
Camshaft drive chain length (see text) .... 309.05 – 309.35 mm (12.167 – 12.179 in)
Service limit .... 311.8 in (12.28 in)

Cylinder head
Maximum warpage .... 0.10 mm (0.004 in)

Crankshaft
Big end bearing axial clearance .... 0.05 – 0.20 mm (0.002 – 0.008 in)
Service limit .... 0.3 mm (0.012 in)
Big end bearing radial clearance .... 0.020 – 0.060 mm (0.0008 – 0.0024 in)
Service limit .... 0.08 mm (0.003 in)
Main bearing radial clearance .... 0.020 – 0.060 mm (0.0008 – 0.0024 in)
Service limit .... 0.08 mm (0.003 in)
Maximum crankshaft run-out .... 0.05 mm (0.002 in)

Primary chain

	750 models	900 model
Type	Hy-Vo	Hy-Vo
Length (see text)	129.78 – 129.98 mm (5.109 – 5.117 in)	139.3 – 139.5 mm (5.484 – 5.492 in)
Service limit	131 mm (5.16 in)	140.9 mm (5.55 in)

Cylinder block

	750 models	900 model
Standard bore size	62.000 – 62.010 mm (2.4409 – 2.4413 in)	64.500 – 64.510 mm (2.5393 – 2.5397 in)
Service limit	62.10 mm (2.445 in)	64.60 in (2.543 in)
Cylinder/piston maximum clearance	0.10 mm (0.004 in)	0.10 mm (0.004 in)
Ovality limit	0.10 mm (0.004 in)	0.10 mm (0.004 in)

Pistons and rings

	750 models	900 model
Piston OD	61.95 – 61.98 mm (2.439 – 2.440 in)	64.46 – 64.49 mm (2.538 – 2.539 in)
Service limit	61.90 mm (2.437 in)	64.40 mm (2.535 in)
Piston ring/groove clearance:		
Top	0.030 – 0.065 mm (0.0012 – 0.0026 in)	0.015 – 0.045 mm (0.0006 – 0.0018 in)
Service limit	0.09 mm (0.004 in)	0.09 mm (0.004 in)
Second	0.025 – 0.055 mm (0.0010 – 0.0022 in)	0.015 – 0.045 mm (0.0006 – 0.0018 in)
Service limit	0.09 mm (0.004 in)	0.09 mm (0.004 in)
Piston ring end gap:		
Top and second rings	0.10 – 0.30 mm (0.004 – 0.012 in)	0.15 – 0.30 mm (0.006 – 0.012 in)
Service limit	0.5 mm (0.020 in)	0.5 mm (0.020 in)
Oil ring (side rail)	0.3 – 0.9 mm (0.012 – 0.035 in)	
Service limit	1.1 mm (0.043 in)	
Gudgeon pin OD	14.994 – 15.000 mm (0.5903 – 0.5906 in)	
Service limit	14.98 mm (0.590 in)	
Gudgeon pin bore	15.002 – 15.008 mm (0.5906 – 0.5909 in)	
Service limit	15.05 mm (0.593 in)	
Gudgeon pin/piston clearance	0.04 mm (0.002 in) max	
Piston/bore clearance	0.10 mm (0.004 in) max	

Clutch

	750 models	900 model
Type	Wet, multiplate	
No. of plates:		
plain	7	
friction	8	
Friction plate thickness:		
Type A	3.72 – 3.88 mm (0.146 – 0.153 in)	
Wear limit	3.4 mm (0.13 in)	
Type B	3.72 – 3.88 mm (0.146 – 0.153 in)	
Wear limit	3.4 mm (0.13 in)	
Plain plate max. warpage	0.3 mm (0.012 in)	
No. of springs	6	
Spring preload at specified length	16.6 – 18.4 kg/ 25.0 mm (36.60 – 40.57 lb/ 0.98 in)	18.3 – 20.1 kg/ 24.4 – 25.6 mm (40.34 – 43.31 lb/ 0.96 – 1.01 in)

Service limit	14.9 kg/25.0 mm (32.85 lb/0.98 in)	17.0 kg/24.4 – 25.6 mm (37.48 lb/0.96 – 1.01 in)
Clutch spring free length	34.2 mm (1.35 in)	35.1 mm (1.38 in)
Service limit	32.8 mm (1.29 in)	33.9 mm (1.33 in)

Gearbox

	750 models	900 model
Type	5-speed constant mesh	
Primary reduction ratio	2.381:1	2.041:1
Gearbox ratios:		
1st	2.533:1	
2nd	1.789:1	
3rd	1.391:1	
4th	1.160:1	
Top	0.964:1	1.000:1
Final reduction	2.533:1	2.588:1
Gear backlash	0.024 – 0.074 mm (0.0009 – 0.0029 in)	0.023 – 0.110 mm (0.0009 – 0.0043 in)
Service limit	0.12 mm (0.005 in)	0.15 mm (0.006 in)
Gearbox pinion bore internal diameter:		
Mainshaft 4th	28.020 – 28.041 mm (1.1031 – 1.1040 in)	
Service limit	28.06 mm (1.105 in)	
Mainshaft 5th	31.025 – 31.050 mm (1.2215 – 1.2224 in)	
Service limit	31.07 mm (1.223 in)	
Layshaft 1st	25.000 – 25.021 mm (0.9843 – 0.9851 in)	
Service limit	25.06 mm (0.987 in)	
Layshaft 3rd	28.020 – 28.041 mm (1.1031 – 1.1040 in)	
Service limit	28.07 mm (1.105 in)	
Mainshaft 5th gear bush OD	30.950 – 30.975 mm (1.2185 – 1.2195 in)	
Service limit	30.93 mm (1.218 in)	
Layshaft 1st gear bush OD	24.959 – 24.980 mm (0.9826 – 0.9835 in)	
Service limit	24.93 mm (0.981 in)	
Layshaft 1st gear bush ID	22.000 – 22.021 mm (0.8661 – 0.8670 in)	
Service limit	22.06 mm (0.869 in)	
Mainshaft OD at 4th gear journal	27.959 – 27.980 mm (1.1007 – 1.1016 in)	
Service limit	27.93 mm (1.100 in)	
Layshaft OD:		
At 1st gear journal	21.987 – 22.000 mm (0.8656 – 0.8661 in)	
Service limit	21.93 mm (0.863 in)	
At 3rd gear journal	27.959 – 27.980 mm (1.1007 – 1.1016 in)	
Service limit	27.93 mm (1.100 in)	
Pinion to shaft/bush maximum clearance:		
Mainshaft 4th/shaft	0.10 mm (0.004 in)	
Mainshaft 5th/bush	0.12 mm (0.005 in)	
Layshaft 1st/bush	0.10 mm (0.004 in)	
Layshaft 1st bush/shaft	0.10 mm (0.004 in)	0.06 mm (0.002 in)
Layshaft 3rd/shaft	0.10 mm (0.004 in)	
Selector fork claw thickness	6.43 – 6.50 mm (0.253 – 0.256 in)	
Service limit	6.1 mm (0.24 in)	
Selector fork bore max ID	13.04 mm (0.513 in)	
Selector fork shaft OD	12.966 – 12.984 mm (0.5104 – 0.5112 in)	
Service limit	12.90 mm (0.508 in)	

Torque settings

Component	kgf m	lbf ft
Cylinder head cover	0.8 – 1.2	6 – 9
Camshaft bearing caps	1.2 – 1.6	9 – 12
Cylinder head	3.6 – 4.0	26 – 29
Camshaft sprocket bolts:		
750 models	2.2 – 2.6	16 – 19
900 models	1.8 – 2.0	13 – 15
Sparking plugs	1.2 – 1.6	9 – 12
Crankcase bolts	See Section 37	
Alternator rotor bolt	8.0 – 10.0	58 – 72
Primary shaft bolt	8.0 – 10.0	58 – 72
Clutch nut	4.5 – 5.5	33 – 40
Gearbox sprocket:		
750 models	5.0 – 5.4	36 – 39
900 models	4.5 – 5.5	33 – 40
Connecting rod nuts:		
750 models	3.0 – 3.4	22 – 25
900 model	3.2	23

Oil filter centre bolt	2.8 – 3.2	20 – 23
Oil pressure switch	1.5 – 2.0	11 – 14
Neutral switch	1.6 – 2.0	11 – 14
Sump drain plug	3.5 – 4.0	25 – 29
Automatic timing unit bolt	3.3 – 3.7	24 – 27
Starter clutch bolts	2.6 – 3.0	19 – 22
Engine mounting bolts	See Section 45	

1 General description

The engine unit forms the most significant aspect of Honda's large capacity four-cylinder range, and is designed to take a now traditional configuration into the 1980s. The units fitted to the various machines in the range are similar in design, having a number of features in common with the six-cylinder CBX engine unit.

The crankshaft is of one-piece construction and runs in five plain bearings. Similar plain big-end bearings support the forged connecting rods. Power from the crankshaft is transmitted through a broad Hy-Vo primary chain to a primary shaft mounted immediately behind the crankcase, where it is transferred through a rubber-and-vane shock absorber to the primary drive pinion and clutch. The right-hand crankshaft end supports the alternator assembly, whilst the starter clutch and CDI ignition reside at the left-hand end of the crankshaft.

Drive to the double overhead camshafts is by Hy-Vo chain from the crankshaft centre to the exhaust camshaft, with a second, smaller, Hy-Vo chain driving the inlet camshaft from the exhaust camshaft. As with most Honda designs, the camshafts are supported directly by the cylinder head material.

Each of the pent-roof combustion chambers carries two inlet and two exhaust valves, which lessens the reciprocating mass of the valve gear and permits greater inlet and exhaust gas flow for a given bore diameter.

Unlike the earlier sohc fours, the new engines feature wet sumps, these being V-shaped to centralise the mass of the oil and to reduce surging. The engine features a high-pressure lubrication system, the feed to the valve gear and camshafts being by means of a separate external feed pipe in preference to the older designs which employed the cylinder head studs as lubrication ducts. The main feed directs oil under pressure to the big-end and main bearing shells, the oil exiting through spray holes in the connecting rods to provide lubrication and cooling for the pistons and bores.

The clutch and gearbox are built in unit with the engine, sharing a common, horizontally-split casing. The clutch is of conventional multi-plate construction, and features one special composite friction plate designed to absorb backlash and transmission snatch by virtue of its resilient design.

The gearbox is of conventional 5-speed, constant mesh construction, the gears running in an oil bath which is shared with the clutch and primary drive.

It will be seen that the new type engine units are rather more than an update of the original four-cylinder unit. The adoption of new designs and materials have resulted in a unit which is both more powerful and cleaner in operation. It follows that the incidence of parts interchangeability between the sohc and dohc is almost negligible, and care must be taken when purchasing replacement parts.

2 Operations with the engine/gearbox unit in the frame

1 It is not necessary to remove the engine unit from the frame to carry out certain operations, although it may prove easier to do so if several tasks need to be undertaken simultaneously. The following components can be removed and/or serviced with the engine in position:

a) Cylinder head cover
b) Camshafts
c) Carburettors
d) Starter motor
e) Gearchange linkage
f) Alternator
g) CDI pickup assembly
h) Clutch

3 Operations with the engine/gearbox unit removed from the frame

1 If attention to the major engine assemblies or internals is required, it will be necessary to remove the unit from the frame and carry out further dismantling on the workbench. Major areas requiring this action are as follows:

a) Cylinder head, block and pistons
b) Crankshaft assembly
c) Primary shaft and primary chain
d) Main and big-end bearings
e) Gear selector mechanism
f) Gearbox selectors and pinions
g) Gearbox bearings and seals
h) Crankcase castings

4 Access to engine/gearbox unit internals

1 The various internal components are housed within the horizontally-split crankcase halves. In order to gain access to these components, it will be necessary to remove the unit from the frame and to separate the crankcase halves, following the procedures detailed in subsequent sections of this Chapter. It follows that the crankcase halves must be reassembled before the unit can be installed in the frame. Crankcase separation will be a necessary precursor to work on items b) to h), listed in Section 3 of this Chapter.

5 Preparing the machine for engine removal

1 Before commencing work on engine removal, the machine should be carefully cleaned down to remove all traces of oil and road dirt. This will make subsequent operations a lot cleaner, and will avoid a lot of unnecessary mess in the workshop. Pay particular attention to the underside of the crankcases and the areas around the various mounting bolts, because these do not normally get cleaned as thoroughly as the rest of the machine.
2 Arrange the machine in the most convenient position in the workshop, allowing as much room as possible around it. Check that the chosen location is well lit, and that the floor area is clean. This will make it easier to retrieve lost nuts or washers should they be dropped. Few owners will possess the luxury of a workshop-type motorcycle ramp, but this can be improvised with a stout wooden bench about 18 inches high, or by using some stout wooden planks supported by crates or concrete blocks. Above all, make sure that any such arrangement is stable, because serious injury can result from a machine toppling over onto the owner. If necessary, the machine may be secured by ropes or tie-down straps for added security.
3 Gather together the necessary tools, some clean rags, and a drain tray or bowl. It is useful to obtain some small boxes so that small related parts can be kept together to aid ident-

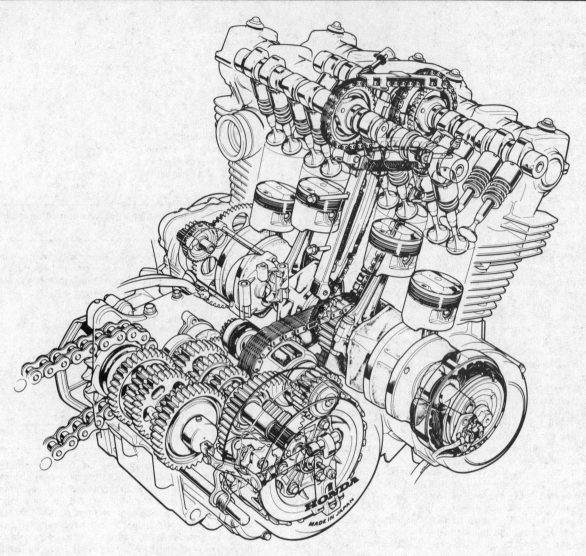

Fig. 1.1 Sectioned view of engine/gearbox unit

ification. Cut-down plastic oil containers are ideal for this purpose, and also make useful drain trays. Before proceeding further, read through the appropriate Sections so that a good general idea of the procedure is gained.

6 Removing the engine/gearbox unit from the frame

1   With the machine set up as described in Section 5, place a bowl or drain tray of about 1 gallon capacity beneath the crankcase drain plug. Remove the plug and leave the oil to drain. If the engine is cold, this will take a considerable time, and can be left to its own devices whilst further dismantling is undertaken.

2   Turn the fuel tap to the off position, and check that there are no naked flames in the vicinity of the motorcycle before prising off the fuel feed pipe at the carburettor.

3   Release the seat latch, and hinge it upwards to gain access to the fuel tank fixing bolt. Slacken and remove the single bolt to free the rear of the tank. The tank can now be lifted and pulled rearwards to free the mounting rubbers at the front. Place the tank to one side in a position where it is not likely to get damaged.

4   The side panels should be released by pulling them free of

their securing rubbers. As a safety precaution, release the battery leads, and preferably remove the battery from the machine to avoid accidental short circuits. Trace the CDI and alternator leads up from the engine, and separate them at the appropriate connection block.

5   Slacken the hose clips which secure the carburettors at the manifold adaptors and to the air cleaner hoses. The carburettor bank can now be disengaged from its mounting stubs and pulled clear. It may prove helpful to release the air cleaner case securing screws so that it can be pushed back clear of the carburettors. Once the carburettors have been moved about halfway out, the throttle cables should be released from the operating quadrant at the centre. This is accomplished by slackening the cable adjuster locknuts and setting the adjusters to give maximum free play. The nipples can then be slid out of the quadrant and the cables removed. The choke cable should be released in a similar fashion, and the overflow and vent pipes guided clear of the frame as the carburettor bank is pulled free.

6   Slacken the nuts which secure the exhaust clamps to their cylinder head studs, allowing the clamps and collet halves to drop clear of the head. The exhaust system can be removed in two sections, after the rear mounting bolts have been released. On the 750 models, the mounting bolts double as footrest securing bolts, whilst on the CB900F, two bolts secure the silencer to the alloy mounting plate. Note that the left-hand

system on CB750K models may prove hard to disengage from the prop stand. The stand should be placed in the down position to aid the removal of the system.

7   The CB900F is equipped with an oil cooler mounted on the front down tubes. Trace the heavy armoured oil pipes down from the cooler to the crankcase unions, releasing the pipes from the engine unit. Take care not to damage the pipes by twisting them as the union bolts are released. Note that a small amount of residual oil will be released, and some provision must be made to catch this.

8   On all models, free the tachometer drive cable from the cylinder head cover by removing the single locking bolt which retains it. The cable should be pulled clear of the engine but need not be removed entirely.

9   Pull off the sparking plug caps, lodging them against the frame top tubes to prevent them from becoming fouled during engine removal. The leads are numbered one to four, number one cylinder being on the left-hand side when viewed from the riding position, and need no further marking. Leave the sparking plugs installed, because this will prevent the ingress of dirt into the cylinders.

10 Because the final drive chain is of the endless type, it will be necessary to slacken it off to permit the removal of the gearbox sprocket. Withdraw the split pin which secures the rear wheel spindle nut. Slacken the nut, and release the chain adjusters so that the wheel can be pushed forwards. If the machine is likely to be moved before the engine unit is refitted, temporarily re-tighten the wheel spindle nut.

11 Moving to the left-hand side of the machine, release the gearchange lever by removing its pinch bolt and drawing it off its splines (US models). European versions, including the UK models, employ a revised footrest position which necessitates the use of a short rear-set linkage between the gearchange lever and selector shaft splines. On these models, the linkage is secured by a pinch bolt to the selector shaft splines, whilst a circlip retains the lever to its pivot on the alloy footrest mounting plate.

12 The UK (European) CB900F, CB750K and the US CB750F feature alloy footrest plates to which the footrests, brake pedal, and on the European models, the gearchange lever are attached. The manufacturers recommend that the left-hand plate is removed to allow sufficient clearance for the left-hand rear engine casing to be removed. In practice this was found not to be essential, because there proved to be enough room to manoeuvre the casing around the plate. The right-hand plate must, however, be released to provide clearance for engine removal. This is secured by the swinging arm pivot shaft nut and the lower engine mounting bolt nut, both of which should be removed temporarily to allow the plate to be withdrawn. On

rear drum brake models, it will be necessary to remove the brake pedal and operating rod, unless the latter is disconnected at the rear wheel and pulled through as the plate is removed. The rear brake switch operating spring must also be disconnected.

13 On rear disc brake models, release the two Allen bolts which secure the rear brake master cylinder to the alloy plate. The brake pedal should also be removed, after releasing its pinch bolt, and the plate lifted away.

14 In the case of the remaining models, the rear brake pedal, gearchange lever and front footrests should be removed, these being attached in a similar manner to that described above, but without the alloy mounting plate. On all models, the left-hand rear engine casing can now be removed after releasing its retaining screws.

15 Slacken the central sprocket retaining bolt and pull the gearbox sprocket clear of its splines. The chain can be disengaged from the sprocket and left to hang against the upper rear mounting bolt whilst the sprocket is placed to one side. This single remaining bolt, which secures the chain guide plate, should be removed, leaving the plate attached to the upper rear mounting bolt alone.

16 Before any attempt is made to release the engine mounting bolts, check carefully to ensure that all electrical and control cables have been disconnected and lodged in a position which will not impede removal. This applies equally to the various drain and breather pipes. The engine is both heavy and bulky, and will require at least two people to remove it in safety. Bearing in mind that the dry weight is around 200 lbs, a jack placed beneath the crankcase is almost essential. Position the jack so that the weight is taken off the mounting bolts, and use a piece of wood between the jack and the delicate sump fins.

17 The engine unit is secured by engine plates bolted to the front of the crankcase and by three long bolts at the rear and underside of the unit. These should be removed, ensuring that the engine remains supported safely by the jack. The front right-hand engine plate also serves to retain the front of the lower frame rail, which is removed to facilitate engine removal. The rear of the removable section is held by two bolts, which should now be released allowing the section to be lifted clear.

18 With one person on each side of the machine, grasp the engine unit firmly, and check that the weight can be taken comfortably before attempting to lift the unit clear. It was found to be advisable to carry out the lifting operation in two stages. First, the unit is lifted through the frame, and the left-hand side rested, whilst the person concerned moves round the frame. The unit can then be swung clear of the frame and placed on the workbench.

6.3 Rear of tank is held by rubber mounted securing bolt

6.4a Remove side panel and disconnect pickup connector (arrowed)

6.4b Disconnect alternator leads at connector block (arrowed)

6.5a Slacken carburettor hose clips (arrowed)

6.5b Release cables as carburettor bank is withdrawn

6.8 Tachometer cable is retained by a single bolt (arrowed)

6.11 Note rearset linkage fitted to UK models

6.12 Remove right-hand footrest plate (where fitted)

6.15 Gearbox sprocket is secured by a single bolt

6.17 Lower frame section can be removed as shown

6.18 Manoeuvre engine unit clear whilst supported by jack

3 An assortment of tools will be required, in addition to those supplied with the machine (see 'Working conditions and Tools' for details). Unlike owners of most Japanese machines, those working on the dohc Honda fours do not suffer unduly from the normal soft-headed cross-point screws. In most areas, small hexagon-headed screws are employed, these being much more robust. In view of this, an appropriate 'nut driver' or box spanner will prove invaluable, in addition to the normal range of workshop tools.

4 Before commencing work, read through the appropriate section so that some idea of the necessary procedure can be gained. When removing the various engine components it should be noted that undue force is seldom required, unless specified. In many cases, a component's reluctance to be removed is indicative of an incorrect approach or removal method. If in any doubt, re-check with the text.

7 Dismantling the engine/gearbox unit: preliminaries

1 Before any dismantling work is undertaken, the external surfaces of the unit should be thoroughly cleaned and degreased. This will prevent the contamination of the engine internals, and will also make working a lot easier and cleaner. A high flash point solvent, such as paraffin (kerosene) can be used, or better still, a proprietary engine degreaser such as Gunk. Use old paintbrushes and toothbrushes to work the solvent into the various recesses of the engine castings. Take care to exclude solvent or water from the electrical components and inlet and exhaust ports. The use of petrol (gasoline) as a cleaning medium should be avoided, because the vapour is potentially explosive and can be toxic if used in a confined space.

2 When clean and dry, arrange the unit on the workbench, leaving a suitable clear area for working. Gather a selection of small containers and plastic bags so that parts can be grouped together in an easily identifiable manner. Some paper and a pen should be on hand to permit notes to be made and labels attached where necessary. A supply of clean rag is also required.

8 Dismantling the engine/gearbox unit: removing the cylinder head cover and camshafts

1 This operation can be tackled with the engine unit installed in the frame or removed for further dismantling. If still in position in the frame, it will be necessary to remove the fuel tank, sparking plug leads and tachometer drive cable before proceeding further. The sparking plugs should be left in position to prevent the ingress of dirt.

2 Remove the eight cylinder head cover retaining bolts, noting that those on the outer edges of the cover differ from the four central bolts. The cover can now be lifted away together with its seal. The seal can be left in position for re-use if it is undamaged.

3 The rear cam chain guide is secured by a total of three bolts, two of which also retain the short oil feed pipe. Remove the bolts and lift away the guide followed by the oil feed pipe and its sealing washers.

4 The camshaft bearing caps are each marked with an identification letter indicating its position on the cylinder head. These letters run from A to L, the letter I being omitted. The cap which incorporates the tachometer drive gearbox is mounted second from the left on the exhaust camshaft and is unmarked. Working from left to right (viewed from riding position) the cap positions and letters are as follows.

Cap position	Marking letter
Ex camshaft LH	*A*
Ex camshaft 1st from left (tachometer)	*Unmarked*
Ex camshaft 2nd from left	*B*
Ex camshaft 3rd from left	*C*
Ex camshaft 4th from left	*D*
Ex camshaft 5th from left	*E*
In camshaft LH	*F*
In camshaft 1st from left	*G*
In camshaft 2nd from left	*H*
In camshaft 3rd from left	*J*
In camshaft 4th from left	*K*
In camshaft 5th from left	*L*

5   When removing the caps, do so in the sequence indicated below. Note that the cap positions can be identified by the corresponding letter cast into the cylinder head immediately below the camshafts. The caps must always be refitted in the correct position.

6   Slacken the bolts which retain caps B, C, H and J. The caps can now be removed and placed to one side. If necessary, use the retaining bolts to dislodge the caps from the camshaft journals by pushing them to one side.

7   Pull off the soft black plastic oil pool caps from the four central cylinder head nuts. Release the rear cam chain guide attachment plate which is located between the two main camshaft sprockets. It is retained by one bolt at this stage, the other having been released when the guide was removed.

8   It will now be necessary to release the pressure from the two cam chain tensioners. Start by slackening the front (exhaust) cam chain tensioner locknut and bolt. This is located at the front of the cylinder head on one side of the cam chain tunnel. Once released, the slipper tensioner will bear upon the short chain connecting the two camshafts. To release pressure, use a large screwdriver to press on the **lower** run of the chain, thus forcing the tensioner back against spring pressure. Hold this position, and secure the tensioner by tightening the adjustment bolt. This should leave the connecting chain with a degree of slack.

9   The tensioner which operates on the crankshaft to exhaust camshaft chain is locked by the **lower** of two small domed nuts located in the centre of the rear face of the cylinder block. Slacken the nut, then moving to the top of the tensioner, visible between the camshaft sprockets, grasp the top section of the tensioner with pliers. Pull this upwards against spring pressure and secure the domed nut. Both cam chains should now be slack.

10   Release the circular inspection cover from the left-hand side of the unit to reveal the CDI pulser assembly. An aperture in the pulser backplate shows an index mark and sequence of timing marks. Using a socket or box spanner, turn the crankshaft until the **1.4T** mark aligns with the index mark. If it proves difficult to turn the crankshaft, loosen or remove the sparking plugs to release cylinder compression. With the crankshaft set as described above, it will be noted that the lobes of both camshafts on No 1 cylinder face inwards, and that the slots in the camshaft end are parallel with the upper jointing face of the cylinder head.

11   Remove the G, K, F and L camshaft bearing caps and place these to one side, together with their locating dowels. The inlet camshaft can now be lifted clear of the cylinder head, disengaged from the connecting chain, and removed.

12   Moving to the exhaust camshaft, one of the two camshaft sprocket securing bolts will now be accessible, and can be removed. Turn the crankshaft anticlockwise until the second bolt appears, and cam lift is at its minimum on any one cylinder. The second bolt can now be released.

13   Release the remaining (D, unmarked, A and E) camshaft bearing caps. The exhaust camshaft can now be disengaged from the sprocket and coupling chain and removed. Release the sprocket from the drive chain, taking care not to allow the chain to fall down into the crankcase. Use a length of wire to retain the chain. It is wise to mark the Hy-Vo chains before removal so that if they are to be re-used they may be refitted to run in the same direction on the sprockets. Reversing the running direction of a partially worn chain is not recommended.

14   If it is wished to leave the valve shimming undisturbed, it is advisable to contrive some means of preventing the shims' escape. This is easily done by cutting two suitable lengths of wooden dowel or tubing to be used as dummy camshafts – an old broom handle being ideal for this purpose. The dummy shafts can then be clamped in place with the camshaft bearing caps until attention to the valves or shims is required. Alternatively, obtain small plastic bags for each of the sixteen valves, marking each bag according to the valve position, and place the cam follower and shim from each valve in the appropreate bag. It is essential that the followers and shims are refitted in their original locations.

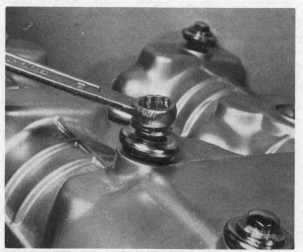

8.2a Remove the rubber-sealed securing bolts ...

8.2b ... and lift the cylinder head cover away

8.3 Guide plate and oil pipe should be removed (bolts arrowed)

8.7 Attachment plate is held by a single bolt

8.14 Dummy camshafts prevent loss of shims

9 Dismantling the engine/gearbox unit: removing the cylinder head

1 The cylinder head can be removed after the engine unit has been removed from the frame (see Section 6). Note that before the head can be removed, it will be necessary to remove the camshafts as described in Section 8 of this Chapter.

2 Release the external oil feed pipe which runs between the intake adaptors of cylinders 3 and 4. The pipe is secured at each end by a hexagon-headed union bolt. Care must be taken not to twist or fracture the oil pipe, and to prevent this the banjo union should be held with an open ended spanner while the bolt is released.

3 Remove the two rear tensioner lock nuts. These are disposed vertically between the 2nd and 3rd cylinders. Moving to the front of the unit, remove the two small bolts which secure the cylinder head to the barrel. These pass upwards into the head on either side of the camchain tunnel.

4 The cylinder head is retained by a total of twelve (12) domed nuts. Starting with one of the nuts nearest to the cam chain tunnel, slacken them in a diagonal sequence, moving each nut by a fraction of a turn to ensure that pressure is released gradually and evenly throughout the cylinder head casting. Failure to observe this precaution can lead to warpage of the cylinder head.

5 The cylinder head can now be lifted clear of the holding studs, whilst the cam chain is fed through the tunnel. If it proves difficult to break the joint between the cylinder block and head, tap gently around the jointing face with a hammer and a hardwood block. On no account attempt to lever the head off, because this will only result in damage to the delicate cooling fins or mating face of the head or block.

10 Dismantling the engine/gearbox unit: removing the cylinder block

1 Before the cylinder block can be removed, it will be necessary to remove the engine unit from the frame and then to dismantle the camshafts and cylinder head. The procedure for these operations is covered in Sections 6 to 9 inclusive. Note that there is insufficient frame clearance to allow the block to be removed with the engine installed in the frame.

2 With the cylinder head removed as described in the previous section, the cam chain tensioner should be pulled clear of the cylinder barrel. It is located by two studs, the external domed nuts having been removed earlier. The cam chain guide at the front of the cylinder block should be lifted clear of its locating groove.

3 The cylinder block will now be retained by a single small bolt which passes down into the crankcase through the cylinder block flange. The bolt should be removed, being located at the front centre of the block.

4 All being well, the block should now be free, and will draw upwards along the holding studs. If it is stuck by the gasket, try tapping around the joint using a hide mallet or a hammer and hardwood block. Take great care not to damage the cooling fins or castings. On no account use a screwdriver or other implement to lever the block free, because this will almost certainly result in damage, leading to oil leakage.

5 As the block comes free, take great care not to allow debris to fall into the crankcase, especially where it is not intended to separate the crankcase halves. As soon as there is room, stuff some clean lint-free rag into each crankcase mouth to catch any carbon, or in dire cases sections of piston ring, which may fall as the pistons are released.

6 The pistons and connecting rods must be supported as they emerge, and prevented from falling against the crankcase. Once clear of all four pistons, the cylinder block can be placed to one side.

9.2 Remove the external oil feed pipe

9.4 Slacken domed nuts in a diagonal sequence

9.5 Lift the cylinder head away

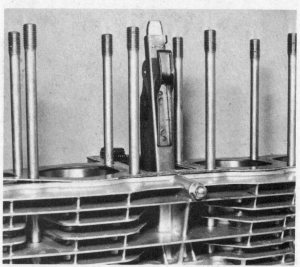

10.2a Release domed nuts and remove tensioner assembly

10.2b Chain guide can be lifted away from tunnel

10.4 Release single bolt, then remove cylinder block

11 Dismantling the engine/gearbox unit: removing the pistons and piston rings

1   Remove the circlips from the pistons by inserting a screwdriver (or a piece of welding rod chamfered one end), through the groove at the rear of the piston. Discard them. Never re-use old circlips during the rebuild.

2   Using a drift of suitable diameter, tap each gudgeon pin out of position, supporting each piston and connecting rod in turn. **Mark each piston inside the skirt so that it is replaced in the appropriate bore.** If the gudgeon pins are a tight fit in the piston bosses, it is advisable to warm the pistons. One way is to soak a rag in very hot water, wring the water out and wrap the rag round the piston very quickly. The resultant expansion should ease the grip of the piston bosses on the steel pins.

3   Do not remove the piston rings at this stage; they should be left in place on the pistons until the examination stage.

12 Dismantling the engine/gearbox unit: removing the clutch assembly

1   The clutch assembly can be removed and replaced irrespective of whether the engine has been removed or is in the frame. No preliminary dismantling operations are necessary to gain access to the clutch or its operating mechanism, other than the disconnection of the clutch operating cable, removal of the rear brake pedal and draining the engine/transmission oil where the unit is installed in the frame.

2   The clutch cover is secured to the crankcase casting by a total of twelve hexagon-headed screws, and can be lifted away after these have been released. If the cover is stuck to the gasket, tap lightly around the joint with a soft faced mallet or a hardwood block and hammer.

3   Slacken the six bolts which retain the clutch release plate in a diagonal sequence, and lift the plate and springs away. The clutch release bearing and pushrod will remain in position in the centre of the plate.

4   The clutch centre is secured by a slotted nut, for which a special Honda tool, No 07716 – 0020100 is available. A similar tool can be made up in the workshop using a short length of thick-walled tubing. Clamp the tube in a vice, and use a hacksaw to cut slots as shown in Fig. 1.2. The shaded area can then be filed away to leave four projecting tangs. These can be used to engage the slots in the nut to facilitate removal.

5   It will be necessary to contrive some means of holding the clutch centre whilst the nut is removed. If the engine is in the frame, select top gear, and apply the rear brake or have an assistant hold the rear wheel. This will hold the centre via the gearbox and final drive chain.

6   If the engine unit is on the bench, an old drive chain can be wrapped around the gearbox sprocket and held with a self-locking wrench. Select top gear and use the wrench to prevent rotation of the clutch centre whilst the nut is removed. An alternative method is to refit the clutch springs with plain washers in place of the release plate. The clutch assembly can then be locked through the primary drive by passing a bar through one of the connecting rod eyes and resting each end on wooden blocks placed on each side of the crankcase mouth. Each method will work equally well. Before slackening the nut, remember to bend back the locking tab which retains it.

7   After the securing nut and tab washer have been removed, the entire clutch assembly can be slid off the end of the gearbox mainshaft without further dismantling. Leave the assembly undisturbed until further examination is required.

12.6 Use spring and plain washers to lock the clutch

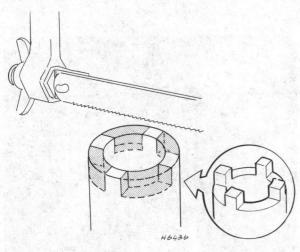
Fig. 1.2 Home-made clutch nut peg spanner

12.7 The entire clutch can now be removed from shaft

13 Dismantling the engine/gearbox unit: removing the primary drive pinion

1   The primary drive pinion is attached to the end of the jackshaft, or primary shaft, and may be removed after dismantling the clutch as described in Section 12 of this Chapter. It should be noted that it is **not** necessary to remove the pinion to facilitate crankcase separation, and unless specific attention to this component is required, this Section may be ignored at this stage.

2   The primary drive pinion is in fact an assembly. The main pinion, having twenty four (24) teeth is splined to the protruding end of the primary shaft. A thin section outer gear fits against the primary drive pinion, and is held against it by pressure from a Belville washer. This outer gear is of the same diameter but has only twenty three (23) teeth, thus for every revolution of the complete assembly, the outer gear is forced to move by one tooth in relation to the primary drive pinion. As a result, any clearance or backlash between the primary gear teeth is taken up by the resistance of the outer gear.

3   To dismantle the above-mentioned assembly, slacken and remove the large Allen bolt which retains it to the primary shaft. Lift away the bolt, followed by the Belville washer. A large plain washer is fitted next, being located by a dowel pin to the primary drive pinion. The outer gear, primary drive gear and spacer can now be slid off the shaft.

14 Dismantling the engine/gearbox unit: removing the alternator assembly

1   The alternator assembly can be removed with the engine unit in or out of the frame. In the case of the former, it will be necessary to trace the braided generator output leads back to the area behind the right-hand side panel, where they should be released at the multi-pin connector.

2   Slacken and remove the three hexagon-headed screws which secure the alternator outer cover to the crankcase. The cover can be lifted away, together with the stator assembly and pickup brush holder.

3   It will be necessary to prevent crankshaft rotation whilst the rotor securing bolt is removed.

4   If the cylinder head and block have been removed, a bar can be passed through one of the connecting rod eyes, the ends resting against wooden blocks placed on each side of the crankcase mouth. Alternatively, if only the alternator requires

attention and the unit is installed in the frame, select top gear and apply the rear brake to lock the entire power train.

5   After the rotor securing bolt has been removed it will be necessary to contrive some means of drawing the rotor off the crankshaft taper. The rotor boss is fitted with a large internal thread which is designed to accept a Honda rotor extractor, part number 07933 – 4250000. If this tool is not to hand, a suitable alternative can be found in the form of the rear wheel spindle. This has the correct thread and a taper which matches that of the crankshaft end. Screw the extractor or spindle into the rotor boss until a fair amount of pressure is applied. If this fails to break the taper joint, tap the head of the extractor or spindle until the rotor springs off. On no account should excessive force be used, nor should the rotor itself be struck.

6   It should be noted that the inner face of the rotor is recessed to provide clearance for the projecting boss which supports the right-hand crankshaft seal. It follows that the rotor must be removed prior to crankcase separation, and that it should not be fitted to the crankshaft end until the crankcase halves have been re-joined. The manufacturer advises that crankcase separation is possible with the rotor in position, but this was found to be impracticable.

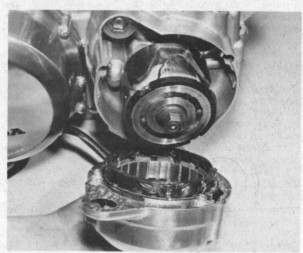

14.2 Remove the alternator cover and stator assembly

14.5a Rear wheel spindle makes effective puller ...

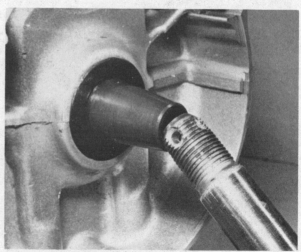

14.5b ... note that taper engages neatly in shaft end

15 Dismantling the engine/gearbox unit: removing the starter motor

1   The starter motor resides in a well formed in the crankcase, immediately to the rear of the cylinder block. The motor can be removed with the engine unit installed in the frame or on the workbench. It is not essential to remove the motor to permit crankcase separation, but it is normal to do so during a full overhaul to permit full inspection of the crankcases and the motor itself. If the motor is to be removed with the engine installed, the following operations should be undertaken as a precursor.

2   Check that the ignition is switched off, then isolate the battery by disconnecting the negative (–) lead. Remove the right-hand side cover, and trace the heavy starter motor cable back to its terminal on the starter solenoid. Disconnect the cable and pull it through to clear the frame.

3   The starter motor is located beneath a pressed steel cover, which is retained by two hexagon-headed bolts. Remove the cover to expose the motor and its two retaining bolts. Release the bolts, then grasp the motor body and pull it to the right-hand side of the engine unit. The motor is sealed by a large O-ring where it enters the starter clutch and CDI pickup housing, and it may prove necessary to lever the motor back against the O-ring's resistance. This must be done very gently to avoid damage. Once freed, the motor can be lifted away, together with its feed cable.

16 Dismantling the engine/gearbox unit: removing the CDI pickup and starter motor clutch

1   The CDI pickup assembly and starter motor clutch are housed beneath the left-hand engine outer cover. The various components may be removed with the engine removed from the frame or in position.

2   Release the left-hand outer cover by removing its eight retaining bolts. Note that it is **not** necessary to remove the circular inspection cover first. The cover will lift away complete with the CDI pickup stator. This need not be disturbed unless specific attention is required, and may be lodged clear of the starter clutch with the wiring intact. If the unit is to be dismantled completely, trace the CDI wiring back from the stator and free it from its guide clips. The cover can now be placed clear of the unit.

3   The CDI reluctor and automatic timing unit (ATU) assembly is secured to the crankshaft end by a single 8 mm bolt. Using a spanner on the flats provided, hold the ATU in position whilst its retaining bolt is removed. The assembly can now be pulled

clear of the crankshaft end. Note the locating pin which engages in the back of the ATU. This precludes any possibility of fitting the unit in the wrong position at a later date.

4   The starter clutch can now be slid off the crankshaft end and placed to one side. It need not be dismantled further unless specific attention is required. An idler gear is fitted between the clutch and starter motor. This takes the form of a double gear pinion, and may be removed together with its support shaft simply by pulling it clear of the casing.

17 Crankcase separation: general information

1   Crankcase separation is possible after the following operations have been carried out. Remove the engine/gearbox unit from the frame (see Section 6). Remove the cylinder head cover and camshafts (see Section 8). Remove the cylinder head (see Section 9). Remove the cylinder block (see Section 10). Remove the pistons (see Section 11). Remove the clutch assembly (see Section 12). Remove the alternator assembly (see Section 14). Remove the starter motor, idler pinion and clutch (see Sections 15 and 16). Remove the CDI pickup (see Section 16).

2   The manufacturer maintains that crankcase separation is feasible with the clutch, alternator and starter motor in position, but **does not** recommend this course of action. It was found that, in the case of the machine used for photographic purposes in this manual, the alternator rotor fouled the crankcase and prevented separation (see paragraph 6, Section 14).

15.3 Starter motor can be removed after releasing bolts

16.2 Remove the outer cover together with the ignition pickup   16.4a Release ATU and lift clutch away

16.4b Idler shaft and pinions can now be withdrawn

18 Dismantling the engine/gearbox unit: separating the crankcase halves

1   Start by removing the eight bolts which pass down from the upper crankcase half. These are all located around the rear section of the crankcase, behind the cylinder block gasket face. The bolts should be slackened in a diagonal sequence, turning each one by about $\frac{1}{2}$ turn until all pressure has been released. This will ensure that no undue pressure is placed on any one area, and will thus prevent any risk of warpage.

2   Turn the unit upside down, supporting it with blocks at the rear to keep it level. The unit can safely be supported by the cylinder block studs at the front, as long as they remain vertical and do not impose any lateral strain on the studs or the casting.

3   Remove the sump after releasing its fourteen holding bolts. Pull off the oil strainer, which is a push fit in the projecting nose of the distributor plate. The plate should be released next, this being retained by a total of six bolts. The separate oil pressure relief valve, located forward of the distributor plate, need not be disturbed at this juncture.

4   The crankcase halves are now secured by a total of twenty four bolts. These should be released in gradual stages, working in a criss-cross pattern to avoid any risk of warpage. Once all the bolts have been released the lower crankcase half can be lifted away. The jointing compound used on the mating surfaces may tend to make separation difficult. If this is the case, tap around the joint with a soft-faced mallet to help break the seal. If necessary, use a hammer and a hardwood block against the more substantial parts of the casing and carefully knock the lower crankcase upwards. Do not use levers between the mating surfaces of the crankcase halves. This will lead to damage of the surfaces and subsequent oil leakage.

19 Dismantling the engine/gearbox unit: removing the gearbox components, crankshaft and primary shaft

1   To gain access to the gearbox components, crankshaft assembly or the primary shaft, it is necessary to remove the engine unit from the frame (see Section 6) and to separate the crankcase halves (see Sections 17 and 18). As the lower crankcase half is removed it will be noted that the selector forks, drum and shaft remain in position, the remaining internal components staying in the inverted upper casing. The gear

selector mechanism can be removed with the engine in the frame, if required, but attention to the selector drum, forks and support shaft will necessitate crankcase separation.

2   Disengage the selector shaft claw from the end of the selector drum by pulling it out against spring pressure. The shaft can now be pulled free of the lower crankcase half. Slacken and remove the shouldered pivot bolt which retains the selector drum detent lever. The return spring should now be released and the lever lifted away. Remove the bolt which retains the index plate to the selector drum end. The plate can be lifted away together with the small selector pins, the spacer and the inner plate.

3   Release the bearing retainer plate which is secured by the location bolt for the selector shaft centring spring, and by a single countersunk screw. The latter will be staked into position and will probably require the use of an impact driver to release it.

4   The selector forks can be released by withdrawing the support shaft from the casing. Note the position of each fork prior to removal, and then refit them to the support shaft in the correct relative order to facilitate reassembly.

5   The selector drum can be withdrawn from the casing, together with its bearing, which is a light push fit in the casing bore. The neutral switch contact blade will remain on the end of the selector drum as it is withdrawn.

6   Moving to the upper casing half, attention can be turned to the gearbox shafts, crankshaft and primary shaft. If the gearbox pinions are suspected of being excessively worn, the backlash between each pair of gears can be measured at this stage, using a dial gauge (DTI) mounted on a suitable stand. Arrange the probe of the dial gauge to rest upon one of the gear teeth, then set the gauge at zero. Rock the pinion back and forth and note the extent of free play (backlash). This should not exceed 0.13 mm (0.005 in) on any pair of pinions. Any excessive clearance is indicative of wear, and may require renewal of the gears concerned.

7   Lift the gearbox mainshaft and layshaft assemblies clear of the casing and place them to one side in their normal operating position until further attention or reassembly is due. The primary shaft should be lifted clear of the casing and disengaged from the primary drive chain. Mark the direction of normal travel of the primary chain before removal so that on reassembly it may be fitted to run in the same direction. Reversing the direction of travel of a partially worn Hy-Vo chain is not recommended. The crankshaft can now be lifted away and placed to one side. Check over both bare crankcase halves, and remove any dowel pins or half-rings which may have been left behind as the various shafts were removed. It is advisable to mark the various small parts to avoid confusion during reassembly.

18.3a Remove the sump from the underside of the unit

18.3b Oil strainer is a push-fit in distributor plate

18.3c Distributor plate is secured by six bolts

19.2a Disengage claw and withdraw selector shaft assembly (arrowed)

19.2b Detent lever is retained by a shouldered bolt (arrowed)

19.2c Dismantle the index plate, pins, spacer and inner plate

19.2d Remove location bolt (A) and screw (B) to release retainer plate

19.4 Withdraw support shaft to free selector forks

19.5 Withdraw selector drum from casing

19.7 Remove gearbox shaft, primary shaft and crankshaft

20 Examination and renovation: general

1   Before examining the parts of the dismantled engine unit for wear it is essential that they should be cleaned thoroughly. Use a petrol/paraffin mix or a high flash-point solvent to remove all traces of old oil and sludge which may have accumulated within the engine.

2   Examine the crankcase castings for cracks or other signs of damage. If a crack is discovered it will require a specialist repair.

3   Examine carefully each part to determine the extent of wear, checking with the tolerance figures listed in the Specifications section of this Chapter. If there is any question of doubt play safe and renew.

4   Use a clean lint free rag for cleaning and drying the various components. This will obviate the risk of small particles obstructing the internal oilways, and causing the lubrication system to fail.

21 Big-end and main bearings: examination and renovation

1   The Honda 750 and 900 dohc models are fitted with renewable shell-type plain bearings on the crankshaft main and big-end journals.

2   Bearing shells are relatively inexpensive and it is prudent to renew the entire set of main bearing shells when the engine is dismantled completely, especially in view of the amount of work which will be necessary at a later date if any of the bearings fail. Always renew the five sets of main bearings together.

3   Wear is usually evident in the form of scuffing or score marks in the bearing surface. It is not possible to polish these marks out in view of the very soft nature of the bearing surface and the increased clearance that will result. If wear of this nature is detected, the crankshaft must be checked for ovality as described in the following section.

4   Failure of the big-end bearings is invariably accompanied by a pronounced knock within the crankcase. The knock will become progressively worse and vibration will also be experienced. It is essential that bearing failure is attended to without delay because if the engine is used in this condition there is a risk of breaking a connecting rod or even the crankshaft, causing more extensive damage.

5   Before the big-end bearings can be examined the bearing caps must be removed from each connecting rod. Each cap is retained by two high tension bolts. Before removal, mark each cap in relation to its connecting rod so that it may be replaced correctly. As with the main bearings, wear will be evident as scuffing or scoring and the bearing shells must be replaced as four complete sets.

6   Replacement bearing shells for either the big-end or main bearings are supplied on a selected fit basis (ie; bearings are selected for correct tolerance to fit the original journal diameter), and it is essential that the parts to be used for renewal are of identical size.

7   Bearing shells should be selected in accordance with the size markings on both the connecting rod and crankshaft. See the following table of sizes:

			CRANKPIN O.D. CODE NO.		
			1	2	3
			35.992–36.000 mm	35.984–35.992 mm	35.975–35.984 mm
CONNECTING ROD I.D. CODE NO.		1 39.000–39.008 mm	E (Yellow)	D (Green)	C (Brown)
		2 39.008–39.016 mm	D (Green)	C (Brown)	B (Black)
		3 39.016–39.024 mm	C (Brown)	B (Black)	A (Blue)

BEARING INSERT THICKNESS:
A (Blue)   : 1.502–1.506 mm (0.0591–0.0593 in)
B (Black) : 1.498–1.502 mm (0.0590–0.0591 in)
C (Brown) : 1.494–1.498 mm (0.0588–0.0590 in)
D (Green) : 1.490–1.494 mm (0.0587–0.0588 in)
E (Yellow): 1.486–1.490 mm (0.0585–0.0587 in)

Fig. 1.3 Big-end bearing shell selection table

8 The relevant crankpin OD (outside diameter) code will be found on the edge of alternate flywheels. In the case of the crankpin, it is the **numeral** (1,2 or 3) that is required, whilst the main bearing journal OD code appears as a **letter** (A, B or C). The crankpin OD code is cross-referenced with the connecting rod code (1, 2 or 3) which is marked across the edge of the big-end eye. The main bearing OD code letter is cross-referenced to corresponding letters stamped at the rear of the crankcase (A, B or C), and will be found in the table below. Note that all crankshaft journals may also be checked by measuring with a micrometer. This method will permit the degree of wear and ovality to be assessed, by comparing the figures obtained with those indicated by the OD codes.

| | | MAIN JOURNAL O.D. CODE NO. | | |
		A	B	C
		35.992–36.000 mm	35.984–35.992 mm	35.975–35.984 mm
CASE I.D. CODE NO.	A 39.000–39.008 mm	D (Yellow)	C (Green)	B (Brown)
	B 39.008–39.016 mm	C (Green)	B (Brown)	A (Black)
	C 39.016–39.024 mm	B (Brown)	A (Black)	E (Blue)

MAIN BEARING INSERT THICKNESS:
A (Black) : 1.498–1.502 mm (0.0590–0.0591 in)
B (Brown) : 1.494–1.498 mm (0.0588–0.0590 in)
C (Green) : 1.490–1.494 mm (0.0587–0.0588 in)
D (Yellow): 1.486–1.490 mm (0.0585–0.0587 in)
E (Blue) : 1.502–1.506 mm (0.0591–0.0593 in)

Fig. 1.4 Main bearing shell selection table

9 The bearing shell thickness for both the main and big-end journals is colour-coded. The shells themselves are marked with a dab of paint on one edge, the colours, and consequently the sizes, corresponding with those given in the tables above.

22 Examination and renovation: crankshaft assembly

1 If the main or big-end bearing shells are found to be worn, the crankshaft journals should be checked with the aid of a micrometer. The journal material will not normally wear at anything like the rate of the soft bearing shells, but if the engine has been run for some time with worn bearings, ovality may develop. The manufacturer does not specify any ovality toler-ance, but as a rough guide, ovality of 0.05 mm (0.002 in) or more will warrant remedial action.

2 The crankshaft should be checked for run-out by supporting it between lathe centres or on V-blocks. Arrange a dial gauge (DTI) to rest upon the centre main bearing journal, then rotate the crankshaft through two complete revolutions, noting the range shown on the gauge. This figure should then be halved to give the actual run-out figure. The service limit for crankshaft run-out is 0.05 mm (0.002 in).

3 The journal surface should be checked carefully for signs of scoring or damage, particularly where badly worn bearing shells have been discovered. If the crankshaft assembly is out of limits or damaged in any way it will be necessary to renew it. The manufacturer does not operate a service exchange scheme or supply undersize bearing shells, so reconditioning by re-grinding the bearing surfaces is not practicable. In some instances, independent engineering companies may be able to re-work a damaged crankshaft.

4 The clearance between any set of bearings and their respective journal may be checked by the use of Plastigauge (press gauge). Plastigauge is a graduated strip of plastic material that can be compressed between two mating surfaces. The resulting width of the material when measured with a micrometer will give the amount of clearance. For example if the clearance in the big-end bearing is to be measured, Plastigauge should be used in the following manner.

5 Cut a strip of Plastigauge to the width across the bearing to be measured. Place the Plastigauge strip across the bearing journal so that it is parallel with the crankshaft. Place the connecting rod complete with its half shell on the journal and then carefully replace the bearing cap complete with half shell onto the connecting rod bolts. Replace and tighten the retaining nuts to the correct torque and then loosen and remove the nuts and the bearing cap. Without bending or pressing the Plastigauge strip, place it at its thickest point between a micrometer and read off the measurement. This will indicate the precise clearance. The original size and wear limit of the crankshaft journals and the standard and service limit clearance between all the bearings is given in the specifications at the beginning of this Chapter.

6 The crankshaft has drilled oil passages which allow oil to be fed under pressure to the working surfaces. Care must be taken to clean these out carefully, preferably by using compressed air.

7 When refitting the connecting rods and shell bearings, note that under no circumstances should the shells be adjusted with a shim, 'scraped in' or the fit 'corrected' by filing the connecting rod and bearing cap or by applying emery cloth to the bearing surface. Treatment such as this will end in disaster; if the

21.5 Remove big-end shells for examination as shown

21.6 The complete big-end bearing assembly

bearing fit is not good, the parts concerned have not been assembled correctly. This advice also applies to the main bearing shells. Use new big-end bolts too – the originals may have stretched and weakened.

8   Oil the bearing surfaces before reassembly takes place and make sure the tags of the bearing shells are located correctly. After the intitial tightening of the connecting rod nuts, check that each connecting rod revolves freely, then tighten to the specified torque setting. Check again that the bearing is quite free.

23 Primary shaft assembly and primary chain: examination and renovation

1   Power from the crankshaft is transmitted to the primary shaft by way of a Morse, or Hy-Vo, chain. The primary shaft is supported at each end by a journal ball bearing, and incorporates a rubber-and-vane type shock absorber. Power from the primary shaft is taken off by a pinion on the right-hand end. This acts as a primary drive pinion, conducting power to the clutch. If it is required that the primary pinion assembly be removed, refer to Section 13 of this Chapter for details.

2   The primary shaft bearings should be carefully washed out with petrol or a suitable cleaning solvent, then checked for wear or roughness by spinning them. Any rough spots or discernible radial play will indicate the need for renewal. The left-hand bearing can be removed after releasing the oil pump pinion circlip, the pinion and its small driving pin, followed by the bearing retaining circlip. The bearing should not prove too tight a fit on the shaft, and may often be removed by jarring the end of the shaft against a hardwood block.

3   The right-hand bearing can be removed after dismantling the primary pinion assembly, as described in Section 13. Do not omit to check the condition of the oil seal which is fitted to the left-hand shaft end. If the sealing lip appears marked or damaged it should be renewed as a matter of course.

4   The primary shaft shock absorber consists of an outer drum with internal vanes which slides over similar vanes on the outside of a splined centre. Rubber damping blocks are fitted between each pair of vanes providing shock absorption in the event of snatch loadings in either direction. The assembly will not normally require attention until the rubber blocks become compressed after very high mileages have been covered.

5   The shock absorber body can be pulled off after the primary pinion and right-hand bearing have been removed. The two halves of the unit are secured by a circlip and plain washer at the right-hand end. Once these have been removed the centre of the unit can be driven out, displacing the damping blocks.

6   Check the rubbers for damage or compression. If the unit is in good condition, there will be no discernible free play when it is assembled, and the rubbers should make the two parts a tight fit. Any slackness will allow snatch loadings to be transferred to the clutch and gearbox, making the machine rather unpleasant to ride. The unit is reassembled in the reverse order of dismantling. If new rubbers prove to be a very tight fit, use a small amount of petrol as a lubricant. This will make assembly much easier, the petrol evaporating off soon afterwards.

7   The Morse type Hy-Vo primary chain is automatically tensioned by a hydraulic tensioner mechanism fed by the engine oil supply. This type of chain is very resistant to stretching, and very high mileages can normally be covered before renewal is necessary. The chain should be checked for wear whenever the engine is stripped for overhaul, following the procedure outlined below. If at or near the service limit, it is worthwhile renewing the chain in view of the considerable amount of dismantling work that will be required should it prove worn out in the near future.

8   Assemble the chain around the crankshaft and primary shaft sprockets, anchoring the crankshaft against suitable stops on the workbench. Attach a spring balance to the primary shaft, and apply a tension of 36 kg (79 lb). With the chain under tension, measure the chain length as shown in Fig. 1.7. The nominal length is 129.78 – 129.98 mm (5.109 – 5.117 in). The chain must be renewed when it reaches the service limit of 131.1 mm (5.16 in).

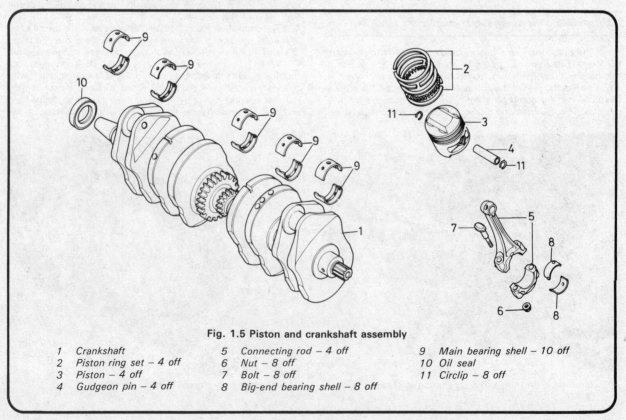

Fig. 1.5 Piston and crankshaft assembly

1 Crankshaft	5 Connecting rod – 4 off	9 Main bearing shell – 10 off
2 Piston ring set – 4 off	6 Nut – 8 off	10 Oil seal
3 Piston – 4 off	7 Bolt – 8 off	11 Circlip – 8 off
4 Gudgeon pin – 4 off	8 Big-end bearing shell – 8 off	

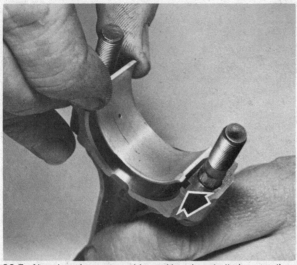

22.7a Note locating tang on big-end bearing shells (arrowed)

22.7b Check that oil drilling is unobstructed (arrowed) ...

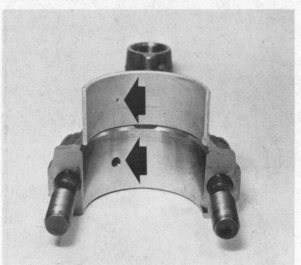

22.7c ... and that holes in con rod and shell align (arrowed)

23.2a Oil pump pinion is secured by circlip

23.2b Lift the pinion clear and displace driving pin

23.2c Bearing retaining circlip can now be released

23.2d Bearing is sliding fit on shaft end

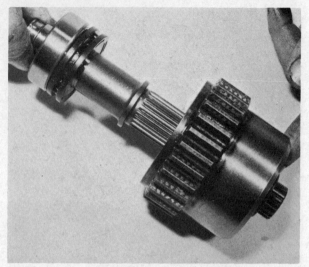
23.5 Shock absorber can be removed from right-hand end

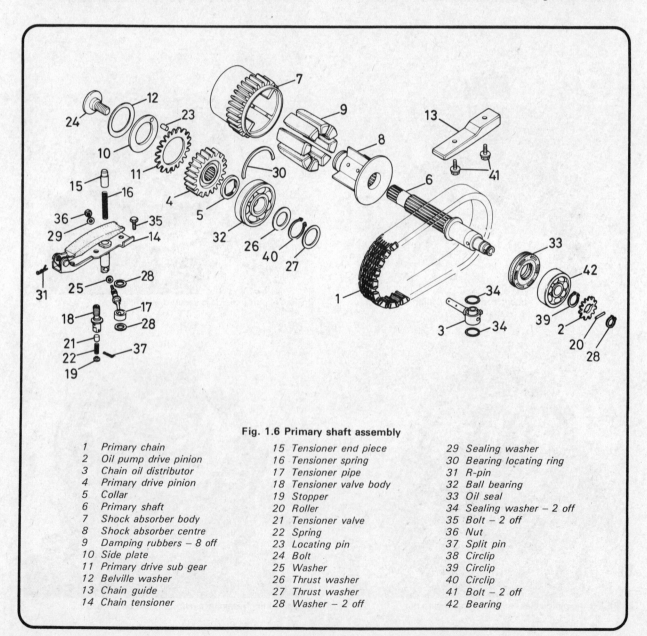

Fig. 1.6 Primary shaft assembly

1 Primary chain	15 Tensioner end piece	29 Sealing washer
2 Oil pump drive pinion	16 Tensioner spring	30 Bearing locating ring
3 Chain oil distributor	17 Tensioner pipe	31 R-pin
4 Primary drive pinion	18 Tensioner valve body	32 Ball bearing
5 Collar	19 Stopper	33 Oil seal
6 Primary shaft	20 Roller	34 Sealing washer – 2 off
7 Shock absorber body	21 Tensioner valve	35 Bolt – 2 off
8 Shock absorber centre	22 Spring	36 Nut
9 Damping rubbers – 8 off	23 Locating pin	37 Split pin
10 Side plate	24 Bolt	38 Circlip
11 Primary drive sub gear	25 Washer	39 Circlip
12 Belville washer	26 Thrust washer	40 Circlip
13 Chain guide	27 Thrust washer	41 Bolt – 2 off
14 Chain tensioner	28 Washer – 2 off	42 Bearing

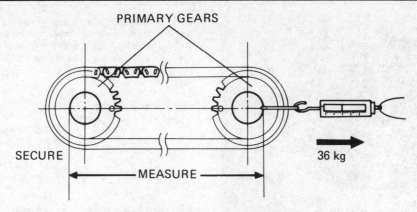

PRIMARY GEARS

SECURE

MEASURE

36 kg

Fig. 1.7 Primary chain wear measurement

24 Oil seals: examination and replacement

1 Oil seal failure is difficult to define precisely. Usually it takes the form of oil showing on the outside of the machine, and there is nothing worse than those unsightly patches of oil on the ground where the machine has been standing. One of the most crucial places to look for an oil leak is behind the gearbox final drive sprocket. The seal should be renewed if there is any sign of a leak.

2 Oil seals are relatively inexpensive, and if the unit is being overhauled it is advisable to renew all the seals as a matter of course. This will preclude any risk of an annoying oil leak developing after the unit has been reinstalled in the frame.

25 Cylinder block: examination and renovation

1 The usual indication of badly worn cylinder bores and pistons is excessive smoking from the exhausts. This usually takes the form of blue haze tending to develop into a white haze as the wear becomes more pronounced.

2 The other indication is piston slap, a form of metallic rattle which occurs when there is little load on the engine. If the top

of the bore is examined carefully, it will be found that there is a ridge on the thrust side, the depth of which will vary according to the rate of wear which has taken place. This marks the limit of travel of the top piston ring.

3 Measure the bore diameter just below the ridge using an internal micrometer, or a dial gauge. Compare the reading you obtain with the reading at the bottom of the cylinder bore, which has not been subjected to any piston wear. If the difference in readings exceeds 0.1 mm (0.004 in) the cylinder block will require boring and honing to the next oversize.

4 If measuring instruments are not available, the amount of bore wear can be approximated as follows. Remove the rings from one piston (see the following Section), then slide it into its bore so that the crown is about ¾ in from the top. Measure the gap between the piston and the bore at 90° to the gudgeon pin boss. If the gap exceeds 0.10 mm (0.004 in) remedial action will be required.

5 If wear has necessitated re-boring, the work should be entrusted to a Honda Service Agent or to a competent engineering shop. Pistons are available in the standard bore size, with oversizes of +0.25, +0.50, +0.75 and +1.00 mm.

6 Make sure the external cooling fins of the cylinder block are free from oil and road dirt, as this can prevent the free flow of air over the engine and cause overheating problems.

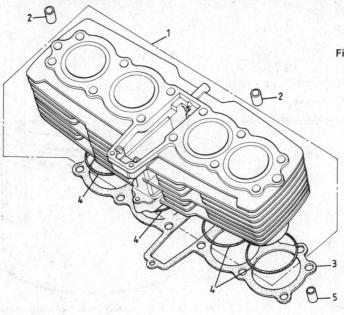

Fig. 1.8 Cylinder block – early CB750K

1   Cylinder barrel
2   Locating dowel – 2 off
3   Cylinder base gasket
4   O-ring – 4 off
5   Locating dowel – 2 off

25.2 Note honing marks on new bore surface. Note O-ring (arrowed)

26 Pistons and piston rings: examination and renovation

1   If a rebore becomes necessary, the existing pistons and piston rings can be discarded because they will have to be replaced by their new oversizes.

2   Remove all traces of carbon from the piston crowns, using a blunt ended scraper to avoid scratching the surface. Finish off by polishing the crowns of each piston with metal polish, so that carbon will not adhere so rapidly in the future. Never use emery cloth on the soft aluminium.

3   Piston wear usually occurs at the skirt or lower end of the piston and takes the form of vertical streaks or score marks on the thrust side of the piston. Damage of this nature will necessitate renewal, if severe.

4   The piston ring grooves may become enlarged in use, allowing the rings to have a greater side float. If the clearance exceeds 0.09 mm (0.004 in) the pistons are due for replacement.

5   To measure the end gap, insert each piston ring into its cylinder bore, using the crown of the bare piston to locate it about 1 inch from the top of the bore. Make sure it is square in the bore and insert a feeler gauge in the end gap of the ring. If the gap is outside the wear limit, the ring(s) concerned must be renewed.

Piston ring end gap

Nominal ring end gap	
Top and 2nd	0.10–0.30 mm (0.004–0.012 in)
Oil control	0.30–0.90 mm (0.012–0.035 in)
Wear limit	
Top and 2nd	0.5 mm (0.020 in)
Oil control	1.1 mm (0.043 in)

6   When refitting new piston rings, it is also necessary to check the end gap. If there is insufficient clearance, the rings will break up in the bore whilst the engine is running and cause extensive damage. The ring gap may be increased by filing the ends of the rings with a fine file, though this is not normally necessary with new rings of Honda manufacture.

7   The ring should be supported on the end as much as possible to avoid breakage when filing, and should be filed square with the end. Remove only a small amount of metal at a time and keep rechecking the clearance in the bore.

8   When dealing with piston rings it is advisable to attend to one piston at a time to preclude the risk of rings being refitted to the wrong piston. This would be undesirable as the rings will have bedded into a particular bore, and would allow compression leakage if fitted incorrectly. It will be noted that the two compression rings differ in that the top ring has a plain profile, whilst the second ring is tapered. Each ring is marked on one face and this should be arranged to face upwards when fitted. The oil control ring is built up from two scraper rails separated by an expander.

9   When installing the piston rings, or when removing sound existing rings for examination, care should be taken not to break them by spreading the ends too far apart. With a little experience the ring gap can be eased apart by hand, just enough to permit removal or fitting. Alternatively, three or four thin metal strips can be placed across the ring grooves and the rings slid on or off. (See the accompanying illustration).

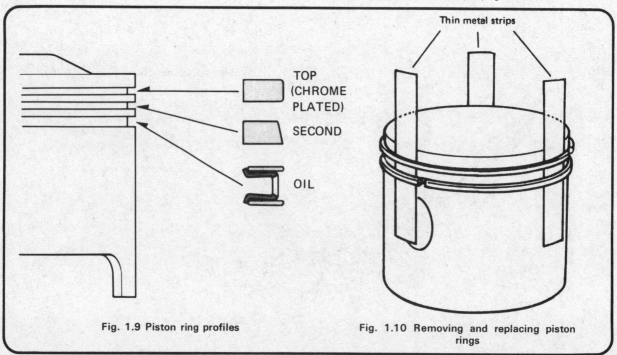

Fig. 1.9 Piston ring profiles

Fig. 1.10 Removing and replacing piston rings

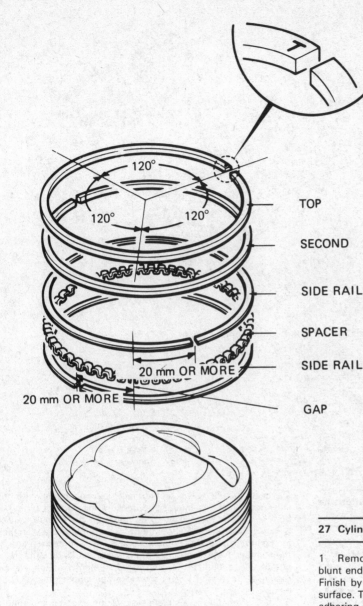

TOP

SECOND

SIDE RAIL

SPACER

SIDE RAIL

GAP

Fig. 1.11 Piston ring end gap positions

26.3 Check piston surface for scoring and wear

27 Cylinder head: examination and renovation

1   Remove all traces of carbon from the cylinder head using a blunt ended scraper (the round end of an old steel rule will do). Finish by polishing with metal polish to give a smooth shiny surface. This will aid gas flow and will also prevent carbon from adhering so firmly in the future.

2   Check the condition of the sparking plug hole threads. If the threads are worn or crossed they can be reclaimed by a Helicoil insert. Most motorcycle dealers operate this service which is very simple, cheap and effective.

3   Clean the cylinder head fins with a wire brush, to prevent overheating, through dirt blocking the fins.

4   Lay the cylinder head on a sheet of $\frac{1}{4}$ inch plate glass to check for distortion. Aluminium alloy cylinder heads distort very easily, especially if the cylinder head bolts are tightened down unevenly. If the amount of distortion is only slight, it is permissible to rub the head down until it is flat once again by wrapping a sheet of very fine emery cloth around the plate glass base and rubbing with a rotary motion.

5   If it proves possible to insert a 0.10 mm (0.004 in) feeler gauge between the glass plate and the cylinder head, the head is beyond the service limit for distortion and remedial action must be taken. If only just outside this limit it may be possible to have the gasket face machined flat, but great care must be taken if this action is chosen. Remember that the machining operation will have some effect on the compression ratio, and might result in the valves touching the piston at high speed. A specialist machinist should be sought for this type of work.

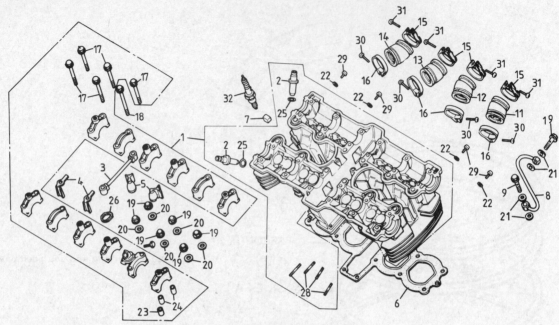

Fig. 1.12 Cylinder head components

1 Cylinder head assembly	12 Carburettor stub	23 Dowel pin
2 Valve guide – 16 off	13 Carburettor stub	24 Dowel pin – 23 off
3 Oil transfer pipe	14 Carburettor stub	25 O-ring – 16 off
4 Oil deflector – 2 off	15 Clip – 4 off	26 O-ring
5 Oil deflector cap	16 Clip – 4 off	27 Bolt
6 Cylinder head gasket	17 Bolt – 22 off	28 Stud – 8 off
7 Rubber damper – 8 off	18 Bolt – 2 off	29 Screw – 4 off
8 Feed pipe	19 Domed nut – 12 off	30 Screw – 4 off
9 Banjo bolt	20 Sealing washer – 12 off	31 Screw – 4 off
10 Banjo bolt	21 Sealing washer – 4 off	32 Sparking plug – 4 off
11 Carburettor stub	22 Thrust washer – 4 off	

28 Valves, valve seats, and valve guides: examination and renovation

1   Remove the cam followers and shims, keeping them separate for installation in their original locations. Compress the valve springs with a valve spring compressor, and remove the split valve collets, also the oil seals from the valve guides, as it is best to renew these latter components. Care should be taken to avoid damage to the cam follower bores when using a valve spring compressor. A Honda service tool is available to protect the soft alloy from scoring (part number 07999 – 4220000). In the absence of this tool a strip of plastic or stout card may be used to line the bore during valve removal. Do not compress the springs more than is necessary to facilitate removal of the split collet halves.

2   As each valve and its associated parts is released, place it in a suitably marked bag or container to ensure that it is refitted in its correct location. If this precaution is not observed, compression leakage will be almost inevitable. When cleaning and examining these components deal with one valve assembly at a time, for the same reason.

3   Each valve should be cleaned prior to checking for wear. Carbon deposits should be removed from the top and underside of the head, taking care not to score the seating face or stem. Remove the heavy carbon deposits using a blunt ended scraper and then finish off with a fine emery paper. When cleaning the stem of the valve use only longitudinal strokes and not rotating strokes. The fine scoring caused by the emery paper may cause stress failure if it runs around the circumference of the valve stem. A highly polished finish is desirable because it reduces the rate of carbon build-up in the future.

4   Examine the valve seating face in conjunction with the valve seat in the cylinder head, looking for signs of pitting or burning. This is most likely to be found on the exhaust valves, because these lead an altogether more strenuous life than their inlet counterparts. If the machine has been maintained properly, there should be no more than a few minor marks on either face, but if severe damage is discovered, remedial action will be required. Small marks can be removed by lapping as described below.

5   Valve grinding is a simple task. Commence by smearing a trace of fine valve compound (carborundum paste) on the valve seat and apply a suction tool to the head of the valve. Oil the valve stem and insert the valve in the guide so that the two surfaces to be ground in make contact with one another. With a semi-rotary motion, grind in the valve head to the seat, using a backward and forward action. Lift the valve occasionally so that the grinding compound is distributed evenly. Repeat the application until an unbroken ring of light grey matt finish is obtained on both valve and seat. This denotes the grinding operation is now complete. Before passing to the next valve, make sure that all traces of the valve grinding compound have been removed from both the valve and its seat and that none has entered the valve guide. If this precaution is not observed, rapid wear will take place due to the highly abrasive nature of the carborundum paste.

6   In view of the number of valves used in these engines, it may be thought worthwhile purchasing one of the oscillatory valve lapping tools which have come onto the market in recent years. This device consists of a sealed gearbox having a driving spindle on one side and a rubber sucker on the other. Rotary movement from an electric drill chuck is converted to the correct to-and-fro motion at the sucker. These devices are well

worth having if more than one or two valves are to be lapped. **On no account** fit the valve stems straight into a drill chuck and attempt grinding by that method, as this will quickly destroy the seat, valve and guide.

7 If a reasonable amount of lapping fails to produce the required unbroken seating area on both the valve and seat, the operation must be abandoned. Excessive lapping is time-consuming and will only result in a ruined valve seat. Examine the seating faces very closely. The attempt at lapping will have highlighted any pits, and a decision must now be taken on the best course of action. Honda advise that the valve should **not** be re-faced, and must therefore be renewed if damaged to the extent that lapping proves ineffective.

8 The valve seats may be re-cut to remove pitting or to compensate for poor seating or incorrect seat widths. Note that if new valve guides are to be fitted, the seats must be re-cut to suit, so check valve guide condition **before** the seats are re-cut. The valve seats are formed by cutting them at three angles to produce the correct seat width, which is nominally 0.99 – 1.27 mm (0.039 – 0.050 in). It should be noted that there is nothing to be gained by using an excessively large contact area, as this will lead to accelerated wear and pitting, and will impair gas flow through the engine.

9 The re-cutting operation requires the use of five separate cutters of various angles and diameters. These are shown in the accompanying diagram (Fig. 1.13), along with the appropriate Honda part numbers. The 32° and 60° cutters are used first, and the correct seating width is then obtained by using the 45° cutter. A word of caution is necessary here, since the valve seats will only accept a limited amount of cutting before they become unacceptably pocketed. When this happens, the seat is no longer usable, and as it is integral with the cylinder head this will require renewal. In view of this risk, and the cost of cutters (not to mention new cylinder heads) it is strongly recommended that any re-cutting is entrusted to a Honda Service Agent. The above information is therefore given for the benefit of owners having access to the required tools and the skill to use them safely.

10 The newly-cut seats should be carefully lapped to their respective valves as described in paragraphs 5 and 6 above. All being well, a perfect gas-tight joint should result. If not, it is likely that the seating face of the valve is unusable, and the valve must be renewed.

11 The amount of valve stem wear must be checked by measuring the stem with a micrometer. Check the stem in a number of positions, and note the smallest reading obtained. If this falls below the service limit, the valve must be renewed.

Valve stem service limit
Inlet	Exhaust
5.47 mm (0.215 in)	*5.44 mm (0.214 in)*

12 The internal diameter of the valve guides can be measured by using ball gauges or a suitable inside micrometer. The service limits for the guides are as follows.

Valve guide ID service limit
Inlet	Exhaust
5.54 mm (0.218 in)	*5.54 mm (0.218 in)*

13 The valve stem diameter figure should now be subtracted from the valve guide ID figure to give the amount of valve to guide clearance. This should not exceed 0.07 mm (0.003 in) for the inlet valve or 0.09 mm (0.004 in) in the case of the exhaust valve. If the total clearance exceeds this, check whether it can be brought within tolerance by fitting a new guide only. Failing this, a new guide and valve must be fitted. In either case, remember that the valve seat must be re-cut.

14 If the correct measuring equipment is not available, check for wear by placing the valve in its guide and attempting to rock the valve to and fro. Anything more than barely discernible play will be indicative of unacceptable wear. If present, take the cylinder head and valves to a Honda Service Agent for verification of wear.

14 The valve guides are an interference fit in the cylinder head, and will require a stepped drift and a certain amount of skill during the renewal operation. The stepped drift should have a spigot of similar size to the valve stem, whilst the larger diameter should be slightly less than that of the valve guide. Support the cylinder head on wooden blocks, so that the combustion chambers face upwards. The valve guide(s) can now be driven out of the head casting. The new component(s) can be fitted in a similar manner, noting that a new O-ring should be fitted beneath the head of each guide prior to its being fitted. Great care should be taken during both operations, as the soft alloy head casting and the guides themselves are easily damaged.

15 The guides must be reamed to size after fitting, using the Honda reamer, part number 07984 – 2000000, or equivalent to bring the ID to 5.500 – 5.515 mm (0.2165 – 0.2171 in). The valve seats must now be re-cut to the new guide as described in paragraphs 8 and 9 above.

16 In view of the amount of skilled work involved in cylinder head reconditioning, some thought should be given to the alternative of entrusting the job to a Honda Service Agent. Bear in mind that some specialist equipment is needed, and is unlikely to warrant purchasing for a one-off job. Much of the cost involved is in the stripping and reassembly work which is a necessary precursor to overhaul. If this stage is undertaken at home, a good proportion of the total cost can be saved.

17 Before reassembling the valves, check the spring seats, springs and collet halves for signs of wear or damage. The valve springs will take a permanent set after very high mileages, and will eventually allow valve float to occur at high speeds. The free lengths of the springs should be measured with a vernier caliper and compared with the table below. Springs are relatively cheap when compared with the cost of rectifying the damage that would result from a valve head hitting a piston at high engine speeds – if in doubt, play safe and renew them.

Valve spring free length — wear limits
Inner	inlet	*39.8 mm (1.57 in)*
	exhaust	*39.8 mm (1.57 in)*
Outer	inlet	*42.5 mm (1.67 in)*
	exhaust	*42.5 mm (1.67 in)*

18 Reassemble the valve and valve springs by reversing the dismantling procedure. Fit new oil seals to each valve guide and oil both the valve stem and the valve guide, prior to reassembly. Take special care to ensure the valve guide oil seal is not damaged when the valve is inserted. On inspection it will be seen that the valve spring coils are more closely wound at one end of the spring than the other. When installing the springs

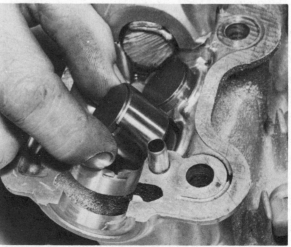

28.1a Remove the cam follower and adjustment shim

28.1b Compress the valve springs and free collet halves

28.1c Remove the springs and spring seat

28.1d Valve can now be displaced and removed

28.1e Check cam follower bores. Note oil seal on guide

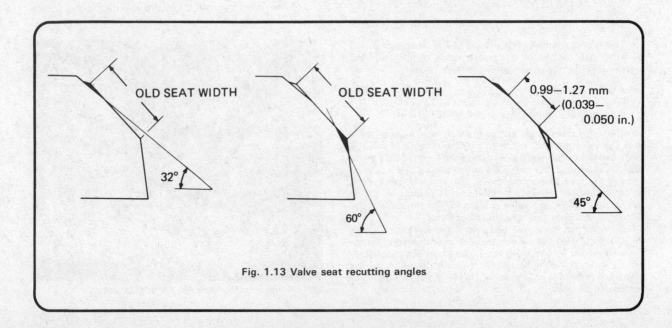

OLD SEAT WIDTH

32°

OLD SEAT WIDTH

60°

0.99–1.27 mm
(0.039–
0.050 in.)

45°

Fig. 1.13 Valve seat recutting angles

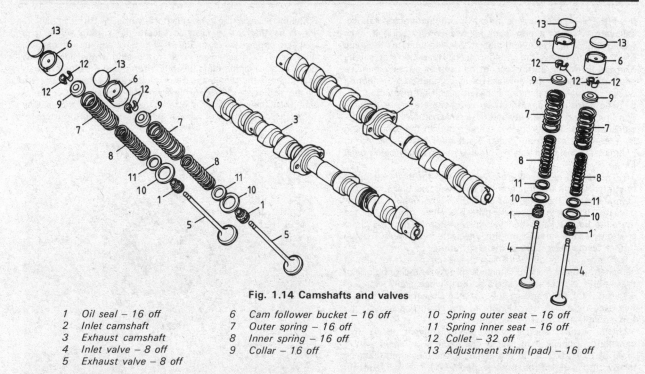

Fig. 1.14 Camshafts and valves

1 Oil seal – 16 off	6 Cam follower bucket – 16 off	10 Spring outer seat – 16 off
2 Inlet camshaft	7 Outer spring – 16 off	11 Spring inner seat – 16 off
3 Exhaust camshaft	8 Inner spring – 16 off	12 Collet – 32 off
4 Inlet valve – 8 off	9 Collar – 16 off	13 Adjustment shim (pad) – 16 off
5 Exhaust valve – 8 off		

ensure that the close wound end faces the cylinder head. As a final check after assembly, give the end of each valve stem a sharp tap with a hammer, to make sure the split collets have located correctly.

29 Camshafts, cam followers and camshaft drive mechanism: examination and renovation

1 Examine the camshaft bearing surfaces for signs of wear or scoring. This type of damage should not occur, because the bearing surfaces are supplied with copious amounts of engine oil during normal running. If, however, the oil has not been changed regularly and the filter has been allowed to become blocked, the bypass system will circulate unfiltered oil carrying abrasive dirt particles to the bearings. The damage that this may cause will be most evident on the soft alloy bearing surfaces in the cylinder head and camshaft caps. Little can be done in these cases, except to renew the cylinder head, bearing caps and the camshafts if these are scored.

2 Camshaft runout can be checked by supporting each end of the camshaft on V-blocks and arranging a dial gauge to bear upon the centre bearing journal. Runout must not exceed 0.05 mm (0.002 in).

3 The clearance between the camshaft bearing surfaces varies along its length according to the table below. It can be measured by using plastigauge in the same way as described in Section 22 of this Chapter.

4 The camshaft and bearing caps should be free from oil prior to measurement. Place a strip of plastigauge on top of each bearing surface, fit the caps in the correct order and tighten down to the recommended torque setting of 1.2 – 1.6 kgf m (9 – 12 lbf ft) in a diagonal pattern to preclude warpage. If the clearance(s) proves to be outside the service limit, check whether renewal of the camshaft(s) will suffice to bring it within limits. Failing this, camshaft and cylinder head renewal will be necessary.

5 Examine each of the camshaft lobes for signs of wear or scoring. A worn camshaft will show signs of flats developing on the peak of each lobe, and the degree of wear can be checked by measuring the lobe at its widest point. The service limit for the inlet camshaft lobes is 36.9 mm (1.45 in) whilst that of the exhaust camshaft lobes is 37.4 mm (1.47 in).

6 The camshaft lobes bear upon hard steel cam followers running in cylindrical bores above the valves and thus to the tops of the valve stems. Clearance between the cam and followers is adjusted by fitting pads of varying thicknesses between. Worn pads would lead to increased valve clearances, but this problem is overcome simply by renewing the pad(s) concerned. The procedure is covered in detail in Routine Maintenance at the front of the manual.

7 There is a small clearance between the cam follower and its bore to allow for lubricating oil. This should not pose a wear problem given regular oil and filter changes, but if wear is evident, the clearance can be checked by measuring the outside diameter of the cam follower and subtracting this reading from the bore measurement.

Camshaft bearing oil clearances

Cap letter code	Nominal	Service limit
A, F, E and L	0.040–0.082 mm (0.0016–0.0032 in)	0.13 mm (0.0051 in)
Unmarked (tachometer drive), G, D and K	0.062–0.109 mm (0.0024–0.0043 in)	0.16 mm (0.0063 in)
B, H, C and J	0.085–0.139 mm (0.0033–0.0055 in)	0.19 mm (0.0075 in)

Cam follower OD

Nomimal	Wear limit
27.972–27.993 mm (1.1013–1.1021 in)	27.96 mm (1.101 in)

Cam follower bore ID

Nominal	Wear limit
28.000–28.016 mm (1.1024–1.1030 in)	28.04 mm (1.104 in)

Cam follower to bore – maximum clearance
0.07 mm (0.003 in)

8   If the clearance exceeds the maximum figure quoted above, calculate whether a new cam follower would bring it back within limits. In the event that this is inadequate, the only solution is to renew the cylinder head. Honda do not supply oversize followers, so this cannot be used as a less expensive alternative unless the services of a competent engineering works are available for the necessary precision machining.

9 ' The camshafts are driven by two separate Hy-Vo chains. The first of these runs between the crankshaft and the exhaust camshaft, the second chain running between the exhaust and inlet camshafts. Spring-loaded tensioners are employed to compensate for chain stretch, a separate assembly being used for each chain.

10 The Hy-Vo type chains are well known for their resistance to stretching, but eventually some wear will take place, with the ultimate result that the tensioner(s) will be unable to compensate for the increased length. The chain lengths should therefore be checked whenever they are removed. If the chain is at or near the service limit it is advisable to renew it to avoid further dismantling work in the near future.

11 Arrange the longer of the two chains (crankshaft to exhaust camshaft) around the two camshaft sprockets, anchoring one of these at the workbench. Using a spring balance, apply a force of 13 kg (29 lb) and measure the distance shown in Fig. 1.16. If this exceeds 311.8 mm (12.28 in) the chain has passed the service limit and must be renewed.

12 The second, shorter chain (inlet camshaft to exhaust camshaft) is checked in exactly the same way, again with a force of 13 kg (29 lb). In this instance, the measured distance (shown in Fig. 1.16) must not exceed 177.1 mm (6.97 in). Do not forget to check the condition of the sprockets, renewing them along with their chains if worn or chipped.

13 The tensioner mechanisms consist of spring-loaded blades which take up any normal chain slack. Tension is set automatically by releasing the tensioner lock bolts when the engine

is idling to allow the tensioner to assume its correct position, the bolts then being used to secure the setting. On a high mileage engine, wear in the cam chains and tensioners will allow a lot of mechanical noise to develop. If the engine was noisy prior to dismantling, despite attempts at tensioner adjustment, check the tensioners' fibre surfaces for wear. If these are deeply scored or damaged they should be renewed in conjunction with their corresponding chain guides. Note that a new tensioner assembly will not compensate for a worn-out chain.

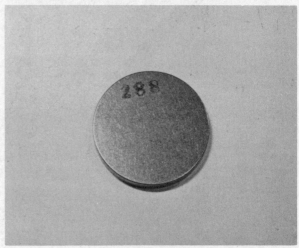

29.6 Cam lobes bear on hard steel adjustment shims

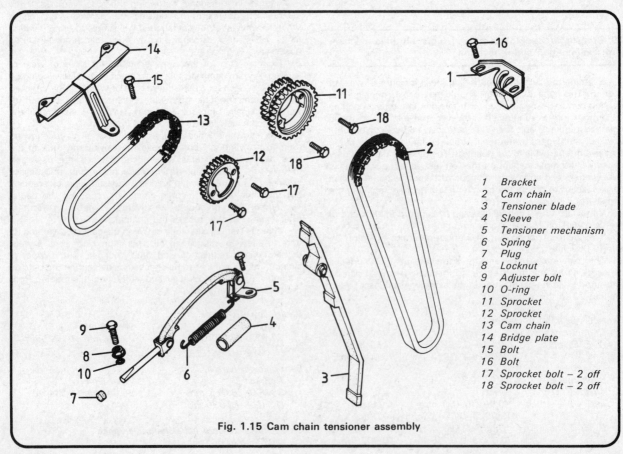

1   Bracket
2   Cam chain
3   Tensioner blade
4   Sleeve
5   Tensioner mechanism
6   Spring
7   Plug
8   Locknut
9   Adjuster bolt
10 O-ring
11 Sprocket
12 Sprocket
13 Cam chain
14 Bridge plate
15 Bolt
16 Bolt
17 Sprocket bolt – 2 off
18 Sprocket bolt – 2 off

Fig. 1.15 Cam chain tensioner assembly

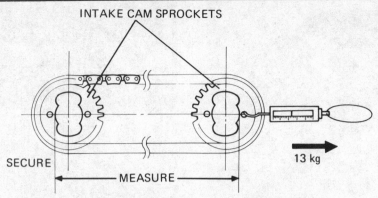

Fig. 1.16 Measuring the cam chain wear

30 Clutch assembly: examination and renovation

1 With the clutch removed as an assembly as described in Section 12 of this Chapter, further dismantling can take place on the workbench. Lift away the flanged clutch centre to expose the first friction plate. This differs from the remainder in having helical slots in the friction material, and must be fitted on the outside of the group with the slots facing the right way. Remove the plain and friction plates which follow. There is now a special spring plate which is in effect two plain plates with a spring arrangement between them to assist freeing and to absorb shocks. Lift away the remaining friction and plain plates, the clutch pressure plate, spacer and splined washer. This will leave the outer drum with its needle roller bearings.

2 After an extended period of service the clutch friction plates will wear sufficiently to allow clutch slip to occur under high loads. The friction plates should be checked for wear by measuring across the lining material using a vernier caliper. The nominal thickness is 3.72 – 3.88 mm (0.146 – 0.153 in). Renew the plate(s) when the wear limit of 3.40 mm (0.13 in) is reached.

3 Clutch slip may also occur on plates within the service limits where the friction surface has become glazed. Glazing is caused by deposits of burnt oil generated by frequent clutch-slipping, and is often found on machines ridden consistently in heavy traffic. It is permissible to remove the glazing by judicious use of abrasive paper, asuming that this does not take the plates below their limit. If this fails to solve the problem, check

the springs as detailed later in this Section.

4 Examine the plain plates for signs of overheating, which may have lead to them assuming a bluish colour. This again is indicative of a heavily-used clutch, and may result in warpage. Place each plain plate on a surface plate or a sheet of plate glass and check for warpage using feeler gauges. If it is possible to insert a 0.3 mm (0.012 in) feeler gauge, the plate(s) must be renewed.

5 Examine the clutch assembly for burrs or indentation on the edges of the protruding tongues of the inserted plates and/or slots worn in the edges of the outer drum with which they engage. Similar wear can occur between the inner tongues of the plain clutch plates and the slots in the clutch inner drum. Wear of this nature will cause clutch drag and slow disengagement during gear changes, since the plates will become trapped and will not free fully when the clutch is withdrawn. A small amount of wear can be corrected by dressing with a fine file; more extensive wear will necessitate renewal of the worn parts.

6 A check of clutch spring condition can be made by measuring the free length. The nominal length is 34.2 mm (1.35 in) and the springs must be renewed when the service limit of 32.8 mm (1.29 in) is reached. If possible, check the spring preload at the specified length, as detailed in the Specifications.

7 The clutch release mechanism consists of a cam-operated lever mounted on the inside of the clutch outer casing, and acting on the clutch release bearing via a mushroom-headed pushrod. The mechanism is relatively sturdy and is unlikely to require attention other than in the case of obvious breakage. The moving parts can be greased whenever the cover is removed for attention to the clutch.

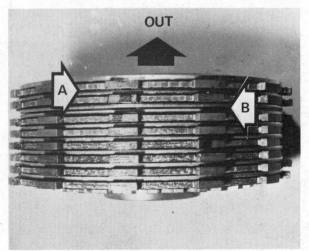

30.1a A: helical-slot friction plate. B: shock absorbing plain plate

30.1b Helical-slot friction plate has extra-wide locating tangs

30.5 Check for wear between plain plates and splines

30.7 The clutch release mechanism. It rarely requires attention

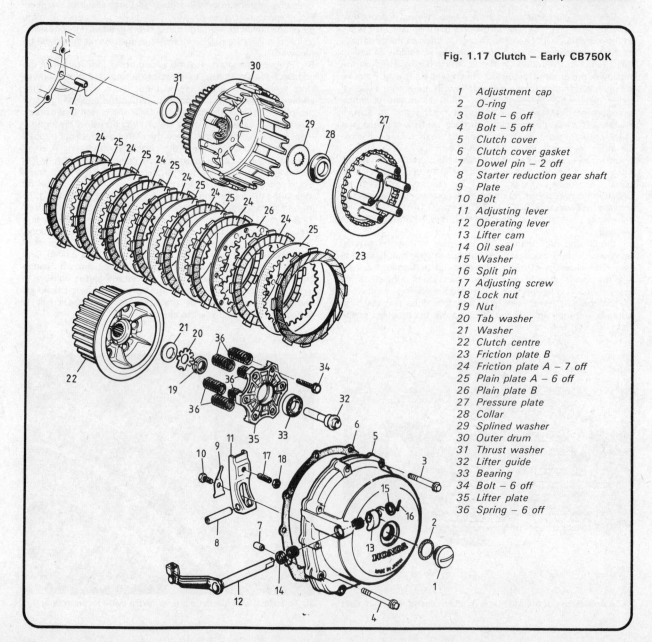

Fig. 1.17 Clutch – Early CB750K

1 Adjustment cap
2 O-ring
3 Bolt – 6 off
4 Bolt – 5 off
5 Clutch cover
6 Clutch cover gasket
7 Dowel pin – 2 off
8 Starter reduction gear shaft
9 Plate
10 Bolt
11 Adjusting lever
12 Operating lever
13 Lifter cam
14 Oil seal
15 Washer
16 Split pin
17 Adjusting screw
18 Lock nut
19 Nut
20 Tab washer
21 Washer
22 Clutch centre
23 Friction plate B
24 Friction plate A – 7 off
25 Plain plate A – 6 off
26 Plain plate B
27 Pressure plate
28 Collar
29 Splined washer
30 Outer drum
31 Thrust washer
32 Lifter guide
33 Bearing
34 Bolt – 6 off
35 Lifter plate
36 Spring – 6 off

31 Gearbox components: examination and renovation

1 Examine each of the gear pinions to ensure that there are no chipped or broken teeth and that the dogs on the end of the pinions are not rounded. Gear pinions with any of these defects must be renewed; there is no satisfactory method of reclaiming them. General wear will lead to backlash developing between the pairs of gears, and this is difficult to check visually. If suspected, check the backlash as described in Section 19.

2 Dismantle each of the gearbox shafts in turn, laying out the components in the order in which they were removed. In the case of the mainshaft (input shaft), the 1st gear pinion is integral, but the remaining gears can be removed after releasing the circlips and washers which secure them. The layshaft (output shaft) can be dismantled in a similar manner, leaving the second gear pinion and bearing in position.

3 Check the internal diameter of each gear pinion, comparing the readings obtained with the service limits listed below. An internal micrometer or vernier caliper will be required for this operation.

Gearbox pinion internal diameters – wear limits
> Mainshaft 4th gear 28.06 mm (1.105 in)
> Mainshaft 5th gear 31.07 mm (1.223 in)
> Layshaft 1st gear   25.06 mm (0.987 in)
> Layshaft 3rd gear   28.07 mm (1.105 in)

4 Check the various gear bushes for wear by measuring the internal diameter (ID) and outside diameter (OD). Renew if beyond the service limit shown below.

Gear bush ID and OD – wear limits
> Mainshaft 5th gear (OD) 30.93 mm (1.218 in)
> Layshaft 1st gear (OD) 24.93 mm (0.981 in)
> Layshaft 1st gear (ID) 22.06 mm (0.869 in)

5 Measure the outside diameters of the mainshaft and layshaft in the positions indicated in Fig. 1.19. Compare the readings obtained with the service limits shown below.

Mainshaft and layshaft OD – wear limits
> A and B 27.93 mm (1.100 in)
> C   21.93 mm (0.863 in)

6 The clearance between each gearbox pinion and the relevant shaft or bush can be calculated from the above. In each case the clearance wear limit is 0.10 mm (0.004 in) with the exception of the mainshaft 5th gear to bush clearance; in this instance the service limit is 0.12 mm (0.005 in).

7 The gearbox bearings should be washed carefully in clean petrol and allowed to dry. The condition of the bearings can be checked by spinning them to highlight any roughness. Any discernible radial play is indicative of the need for renewal. Very slight axial play is normal and thus acceptable, but if excessive will require renewal. Inspect the bearing tracks and the balls or rollers. These will be unmarked and highly polished in a sound bearing, any surface defects indicate that the bearing is worn out. Retain any sound bearing, and lubricate it with oil to prevent corrosion forming prior to reassembly.

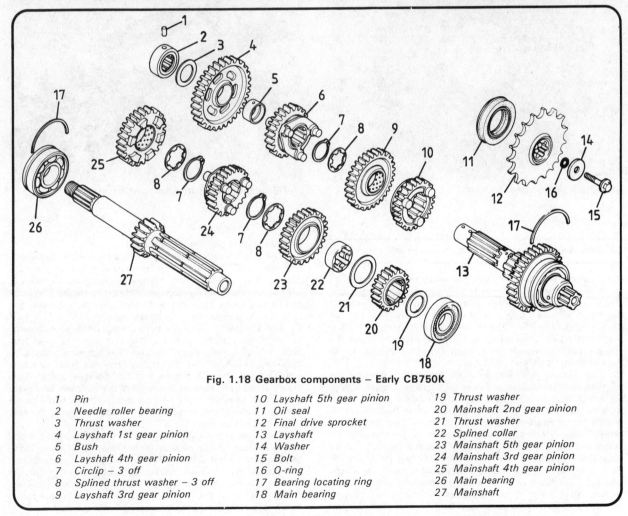

Fig. 1.18 Gearbox components – Early CB750K

1 Pin	10 Layshaft 5th gear pinion	19 Thrust washer
2 Needle roller bearing	11 Oil seal	20 Mainshaft 2nd gear pinion
3 Thrust washer	12 Final drive sprocket	21 Thrust washer
4 Layshaft 1st gear pinion	13 Layshaft	22 Splined collar
5 Bush	14 Washer	23 Mainshaft 5th gear pinion
6 Layshaft 4th gear pinion	15 Bolt	24 Mainshaft 3rd gear pinion
7 Circlip – 3 off	16 O-ring	25 Mainshaft 4th gear pinion
8 Splined thrust washer – 3 off	17 Bearing locating ring	26 Main bearing
9 Layshaft 3rd gear pinion	18 Main bearing	27 Mainshaft

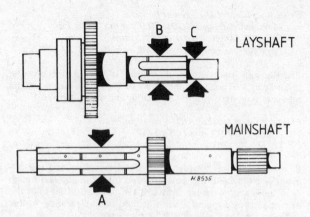

Fig. 1.19 Measuring gearbox shaft wear
Refer to text for details

31.1 Check pinion teeth for wear or chipping

31.4 Check pinion bushes for wear

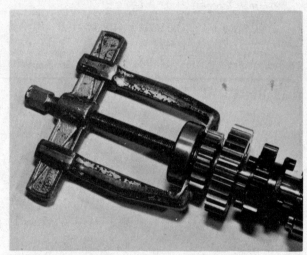

31.7 Bearings may require puller to effect removal

32 Gear selector mechanism: examination and renovation

1   Examine the selector forks for wear, noting the surface finish around the fork ends where they engage with the gearbox pinions. Wear here is unlikely unless lubrication has been badly neglected, in which case scoring may be evident. The width of the fork should be measured for wear, renewal being necessary if the reading obtained is less than 6.1 mm (0.24 in). Check the internal bore measurement of each fork. This should not be greater than 13.04 mm (0.573 in).

2   The selector fork support shaft should be examined for signs of wear or scoring, and rejected if badly damaged. Check for straightness by rolling the shaft on a dead flat surface, such as a surface plate or a sheet of glass. If at all bent, the shaft should be renewed as it will seriously impair gear selection if refitted. The outside diameter of the shaft can be measured for wear if there is some doubt as to its condition. It must be renewed if worn to 12.90 mm (0.508 in) at any point. Check the fit of the shaft in its casing bore. It should normally be a light sliding fit. Excessive wear will allow movement, and thus sloppy gearchanging action. If the casing has become worn, it may be necessary to have it bored out and bushed. This should be left

to a competent engineering company.

3   The selector drum is supported by a journal ball bearing at one end, the other end running directly in the casing. The bearing does not lead a demanding existence, and is unlikely to warrant attention during the normal life of the engine. The same can be said of the plain end, and neither will require more than a cursory inspection for wear or damage.

4   The selector drum tracks, on the other hand, are subjected to fairly high loadings at times, and may begin to wear after high mileages have been covered. The grooves should be examined in conjunction with the selector fork guide pins with which they engage. It is normal to find polished areas where pressure has been exerted, but most of the wear will take place on the comparatively small selector fork guide pin. It will be the latter component that is most likely to require renewal in cases where wear is severe.

5   The selector claw assembly conveys movement to the end of the selector drum by way of hardened steel pins. These components do not suffer from wear in normal circumstances, but should be checked, paying particular attention to the working surfaces. If the machine has shown a tendency to jump out of gear, check the detent mechanism, renewing the spring as a precaution.

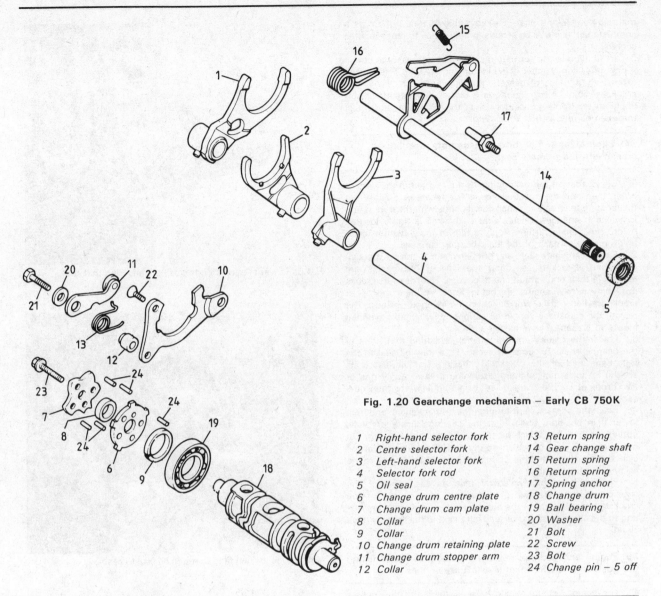

Fig. 1.20 Gearchange mechanism – Early CB 750K

1	Right-hand selector fork	13	Return spring
2	Centre selector fork	14	Gear change shaft
3	Left-hand selector fork	15	Return spring
4	Selector fork rod	16	Return spring
5	Oil seal	17	Spring anchor
6	Change drum centre plate	18	Change drum
7	Change drum cam plate	19	Ball bearing
8	Collar	20	Washer
9	Collar	21	Bolt
10	Change drum retaining plate	22	Screw
11	Change drum stopper arm	23	Bolt
12	Collar	24	Change pin – 5 off

32.5 Examine selector claws for wear or damage

33 Engine reassembly: general

1 Before reassembly of the engine/gear unit is commenced, the various component parts should be cleaned thoroughly and placed on a sheet of clean paper, close to the working area.

2 Make sure all traces of old gaskets have been removed and that the mating surfaces are clean and undamaged. One of the best ways to remove old gasket cement is to apply a rag soaked in methylated spirit or where necessary, a gasket cement solvent. This softens the cement allowing it to be removed without resort to scraping and the consequent risk of damage.

3 Gather together all the necessary tools and have available an oil can filled with clean engine oil. Make sure all new gaskets and oil seals are to hand, also all replacement parts required. Nothing is more frustrating than having to stop in the middle of a reassembly sequence because a vital gasket or replacement has been overlooked.

4 Make sure that the reassembly area is clean and that there is adequate working space. Refer to the torque and clearance settings whenever they are given. Many of the smaller bolts are easily sheared if over-tightened. Always use the correct size screwdriver bit for the crosshead screws and never an ordinary screwdriver or punch. If the existing screws show evidence of

maltreatment in the past, it is advisable to renew them as a complete set, using Allen screws in preference to cross-headed screws.

5   In addition to the above items, it will be necessary to obtain a tube of silicone rubber (RTV) jointing compound. This is used extensively in place of gaskets, particularly in the case of the crankcase joint. Also required is a can of molybdenum disulphide grease (Moly grease) for use as the initial lubricant whilst the main oil feed is first circulating.

34 Engine and gearbox reassembly: replacing the crankshaft and primary shaft

1   Check that all the bearing shells are laid out in the correct order, then refit them to their respective recesses. Ensure that the locating tang on each shell corresponds with the depression in which it engages. Ensure that each shell is firmly located before proceeding further. Apply a film of molybdenum disulphide grease to each of the main bearing surfaces.

2   Fit the camshaft chain and primary chain to their respective crankshaft sprockets, ensuring that the Hy-Vo chains are refitted in their original running directions. This, of course, does not apply if new chains are being fitted. Arrange the upper crankcase half on the workbench in the inverted position. The rear of the crankcase should be supported by suitable wooden blocks to present a level gasket surface.

3   Lower the crankshaft into position, ensuring that none of the bearing shells become displaced. Fit a new oil seal to the right-hand end of the crankshaft, having first lubricated its sealing lip with molybdenum disulphide grease. Apply one or two drops of Loctite or a similar sealant/adhesive to retain the seal.

4   Place the primary shaft through the primary chain, and then lower it in position. Check that the bearings locate properly, noting the half-ring which locates the right-hand bearing. **Do not** allow any form of sealant to get onto the primary shaft oil seal; this acts as part of the lubrication system, feeding oil to the primary shaft assembly.

5   Fit the large hollow dowel pins at each end of the crankcase. There are two of these; one at each end of the crankshaft, to the rear of the outer main bearings. Fit a new O-ring to the chain tensioner oil feed joint next to the primary drive chain.

35 Engine and gearbox reassembly: rebuilding and fitting the gearbox mainshaft and layshaft assemblies

1   The gearbox shafts should be reassembled in the reverse order of the dismantling sequence. Reference should be made to the exploded view of the assembly and to the accompanying photographic sequence for details of the disposition of the various components. Although reassembly is generally straightforward, the following points should be noted.

2   When fitting the mainshaft 5th gear bush, note that its oil hole must align with that of the mainshaft. A similar arrangement will be found in the case of the layshaft 5th gear pinion. An oil hole is provided to permit a lubrication feed from the shaft to the selector fork groove. Ensure that the holes align correctly. Check that the circlips seat squarely in their grooves, and that they are not slack when fitted. If in doubt, fit new circlips to preclude failure at a later date.

3   Fit a half-ring in the bearing support boss groove of the right-hand end of the mainshaft. A half-ring should also be fitted to the inner groove of the left-hand layshaft bearing boss. Note that the outer groove is provided as a means of location for the oil seal. Fit the dowel pin in position in the right-hand layshaft bearing boss.

4   Fit a new, greased, oil seal to the right-hand end of the layshaft. The two assembled shafts can now be fitted into the casing. Check that the half-rings, dowel pin and oil seal locate properly, and that the gearbox pinions turn smoothly and evenly.

34.3a Lower crankshaft into position, ensuring that seal engages

34.3b Use locking fluid to retain oil seal to casing

34.5 Fit new O-ring to oil feed joint (arrowed)

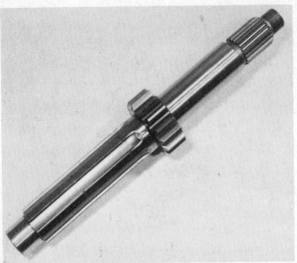

35.1a The gearbox mainshaft and 1st gear pinion

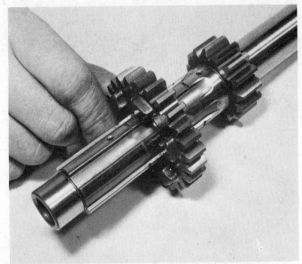

35.1b Fit the 4th gear pinion with dogs facing as shown

35.1c Fit the splined thrust washer and secure with circlip

35.1d Fit 3rd gear pinion, noting that selector groove is fitted innermost

35.1e Place circlip in groove as shown ...

35.1f ... followed by splined thrust washer

35.1g Fit 5th gear bush, ensuring that oil holes align

35.1h Fit the 5th gear pinion ...

35.1i ... followed by thrust washer and 2nd gear pinion

35.1j Small diameter thrust washer is fitted as shown

35.1k Bearing is fitted with sealed face outwards

35.1l Fit bearing and thrust washer to clutch end

35.2a The gearbox layshaft and 2nd gear pinion

35.2b Fit 5th gear pinion, aligning oil holes (arrowed)

35.2c Fit 3rd gear pinion as shown

35.2d Fit splined thrust washer ...

35.2e ... and secure with circlip

35.2f Fit 4th gear pinion, selector groove inwards

35.2g Fit bush and layshaft 1st gear pinion

35.2h Needle roller bearing is fitted as shown

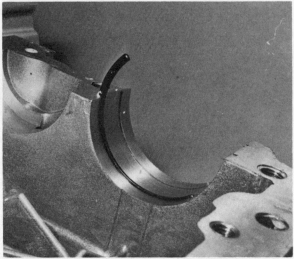

35.3a Fit half ring to right-hand mainshaft and left-hand layshaft bearing bosses

35.5b Right-hand layshaft bearing is located by dowel pin

35.4a Place the assembled layshaft in position ...

35.4b ... followed by the mainshaft assembly

36 Engine and gearbox reassembly: refitting the gear selector mechanism and primary chain tensioner

1   Fit the neutral switch contact to the end of the selector drum, then slide it into position via the larger bearing boss. Fit the selector fork support shaft, placing each of the three forks in position as the shaft is slid into place. Note that the forks differ in shape, and should be arranged as shown in Fig. 1.20.

2   Position the selector drum bearing retainer and secure it with a countersunk cross-head screw. This goes at the top right-hand corner when viewed from the end of the drum. The extreme left-hand end of the retainer is held by the selector claw centring spring anchor pin. This has a hexagon head with a small extension to engage with the spring ends. Note that a thread locking fluid should be used on both screws prior to fitting, and that the countersunk screw should be staked in position after tightening.

3   Fit the spacer to the end of the selector drum, then position the inner index plate. The latter has a number of holes, one of which is much closer to the centre than the remainder. This should engage with the locating pin in the end of the drum. Fit the four pins to the remaining holes, then offer up the outer index plate. Note that this cannot be fitted incorrectly because the pins are grouped asymetrically. Fit the central retaining bolt and tighten down firmly.

4   Assemble the detent lever, spacer and hairpin spring on the pivot bolt, then offer this up to the casing. The bolt occupies the remaining bearing retainer thread and serves as a means of retaining it and the detent assembly. Check that the detent lever roller engages with the cam profile on the outer index plate and that it is under spring pressure. Tighten the bolt to 1.0 – 1.4 kgf m (7 – 10 lbf ft).

5   Lubricate the selector drum tracks with engine oil, then check the assembly for free operation. Note that neutral can be found by positioning the drum so that the detent lever engages the smallest of the six depressions in the outer index plate. Check that the neutral switch contact is touching the switch contact pin.

6   Check that the gearchange lever shaft oil seal is in good condition. If in any doubt, renew it as a precaution against leakage. The seal lips should be coated with grease prior to fitting the shaft. Take care not to damage the seal with the shaft splines during installation. It is worth wrapping the splines with PVC tape to protect the seal at this stage. Slide the shaft into position, lifting the claw mechanism over the end of the index plate. Check that the centring spring engages with its anchor pin. Temporarily refit the gearchange lever and check that all five gear positions plus neutral can be selected.

7   If the primary chain tensioner has been removed for examination, it must be refitted at this stage. Place the tensioner assembly in position in the casing and fit its three retaining bolts. Fit the spring and plunger into their housing, then hinge the tensioner blade into place. On the outside of the casing, assemble the oil feed pipe and union to the top of the tensioner body. Refit the tensioner oil valve, taking care not to fracture the short oil feed pipe as it is tightened. It is important that the internal components of the tensioner body and oil valve are kept clinically clean during reassembly.

37 Engine and gearbox reassembly: joining the crankcase halves

1   Clean the crankcase mating faces with high flash-point solvent or methylated spirit to remove any residual grease spots or dust. Coat the lower casing half mating face with a silicone rubber (RTV) jointing compound, taking care not to obstruct any oilways.

2   **Important note:** On no account allow the jointing compound to get near the main bearing shells. A border of about $\frac{1}{8}$ in must be left clear to prevent the compound finding its way onto the bearing surface when the casing halves are joined. If this precaution is not observed the bearing oil feeds may become blocked, resulting in seizure. Similar care should be exercised when applying jointing compound in the vicinity of the oil jet in the lower casing half. Refer to Fig. 1.23 for details of jointing compound application.

3   Set the gear selector mechanism in the 1st gear position, and set the gearbox components to 1st gear by sliding the layshaft 4th gear into engagement with the layshaft 1st gear. The remaining gears should be left disengaged. Lubricate the mainshaft 3rd gear selector groove and each of the main bearings with molybdenum disulphide grease.

4   Carefully lower the lower crankcase half onto the inverted upper half, ensuring that the gear selector forks engage in their grooves. It was found that it is necessary to restrain the primary chain tensioner blade as the casing was lowered. This was accomplished by inserting a finger through the hole in the casing, and it proved possible to prevent the blade pivoting down, thus freeing the plunger and spring. For those possessing index fingers of the wrong shape or size, a piece of stiff wire bent at 90° at the end will do quite well. As soon as the tensioner blade is in contact with the primary chain it can be released.

5   The majority of the crankcase holding bolts are fitted from the underside of the unit, and these should be dropped into position. The ten bolts which also retain the crankshaft assembly should each have a plain washer fitted. Note that the threads of these bolts should be lubricated with a smear of molybdenum disulphide grease. The above-mentioned bolts, like most of the remainder are 8 mm, whilst a row of 6 mm bolts are fitted along the front edge. A single 10 mm bolt is located at the rear of the casing. Note the wiring clip which should be fitted beneath the head of the No 24 bolt.

6   Refer to the tightening sequence shown in the accompanying illustration, and tighten each bolt down to about $\frac{1}{2} - \frac{2}{3}$ of its recommended torque setting. Then go over each one again to bring it up to the full torque value. The torque settings are as shown below.

Crankcase torque settings
All   6 mm bolts: 1.0 – 1.4 kgf m (7 – 10 lbf ft)
All   8 mm bolts: 2.1 – 2.5 kgf m (15 – 18 lbf ft)
All 10 mm bolts: 4.5 – 5.0 kgf m (33 – 36 lbf ft)

36.1 Fit neutral switch contact (arrowed) then fit selector drum

36.6 Renew seal if worn. Lubricate before fitting shaft

36.7a Assemble tensioner spring and plunger

36.7b Check that retaining clip is secure (arrowed)

37.4 Hold tensioner blade in position as casing is lowered

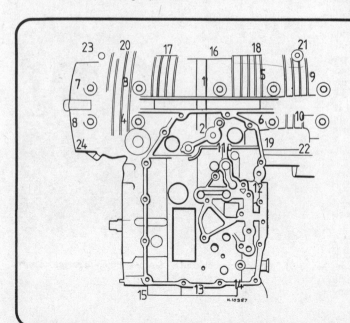

Fig. 1.21 Crankcase bolt tightening sequence

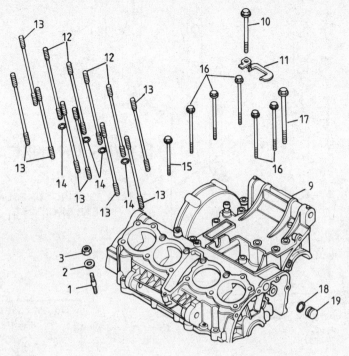

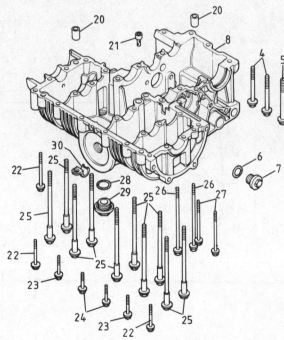

1   Stud
2   Washer
3   Nut
4   Bolt – 2 off
5   Bolt
6   Sealing washer
7   Oilway plug
8   Lower crankcase
9   Upper crankcase
10 Bolt – 3 off
11 Shield plate
12 Stud – 4 off
13 Stud – 8 off
14 O-ring – 4 off
15 Bolt – 4 off
16 Bolt – 5 off
17 Bolt
18 O-ring
19 Plug
20 Dowel pin – 2 off
21 Oil control piece
22 Bolt – 4 off
23 Bolt – 2 off
24 Bolt – 2 off
25 Bolt – 10 off
26 Bolt – 3 off
27 Bolt – 2 off
28 O-ring
29 Drain plug
30 Cable clip

Fig. 1.22 Crankcases – Early CB750K

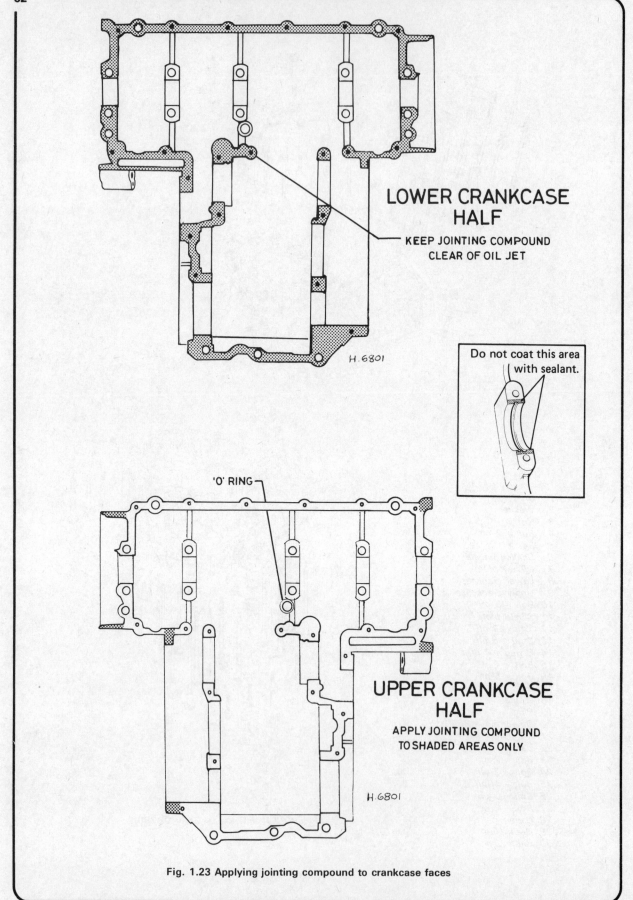

LOWER CRANKCASE HALF

KEEP JOINTING COMPOUND CLEAR OF OIL JET

H.6801

Do not coat this area with sealant.

'O' RING

UPPER CRANKCASE HALF

APPLY JOINTING COMPOUND TO SHADED AREAS ONLY

H.6801

Fig. 1.23 Applying jointing compound to crankcase faces

38 Engine and gearbox reassembly: refitting the oil distributor plate and sump

1   Place the oil distributor plate on the underside of the crankcase and fit its six retaining bolts. Refit the oil strainer assembly, using a new sealing ring where required. Clean the sump (oil pan) gasket face, taking care not to scratch the surface. Position a new sump gasket, then offer up the sump itself. Tighten the 14 retaining bolts in a diagonal sequence to avoid any risk of distortion.

39 Engine and gearbox reassembly: refitting the primary drive pinion and clutch assembly

1   If the primary drive pinion assembly was removed during the engine overhaul it must be refitted **before** the clutch is installed. Fit the spacer to the end of the primary shaft, followed by the pinion, ensuring that the small locating dowel faces outwards. Assemble the large plain washer, noting that it should engage with the locating pin, followed by the Belville washer. Lock the crankshaft by the same method that was chosen during removal, then fit and tighten the large-headed Allen bolt. The bolt should be pulled down very tight. If a torque wrench is available with the appropriate hexagon key, a torque setting of 8.0 – 10.0 kgf m (60 – 72 lbf ft) should be used.
2   Place the plain thrustwasher over the projecting gearbox

mainshaft end, followed by the clutch outer drum. It will be necessary to turn the crankshaft until the area of the primary drive pinion where the teeth on the inner and outer sections are in synchronisation meshes with the clutch drum's driven gear. Check that the caged needle roller bearings which support the drum are in position and are lubricated.
3   Fit the splined washer, followed by the spacing collar, noting that its smaller diameter faces outward. If the clutch pressure plate, centre and clutch plates are still assembled, they can now be fitted as a unit. Alternatively, proceed as follows.
4   Place the clutch friction and plain plates over the splined clutch centre, noting that the helically-grooved friction plate should be fitted first, with the groove inclined as shown in Fig. 1.17. This is followed by normal plain and friction plates, the second plain plate being the composite shock-absorbing plate described earlier in the Chapter. Refer to Fig. 1.17 for details of the assembly order. Note that it is almost impossible to assemble the clutch plates in the drum prior to fitting the clutch centre, as this would make alignment of the internal splines of the plain plates difficult.
5   Place the plain washer and tab washer in position, then fit the clutch centre securing nut finger tight. Using the method employed during dismantling, lock the clutch centre and tighten the slotted securing nut to a torque setting of 4.5 – 5.5 kgf m (33 – 40 lbf ft). Do not forget to bend up the tab washer.
6   Place the six clutch springs in position and offer up the clutch release plate. Fit the bolts in a diagonal sequence, tightening each one in turn to prevent distortion.

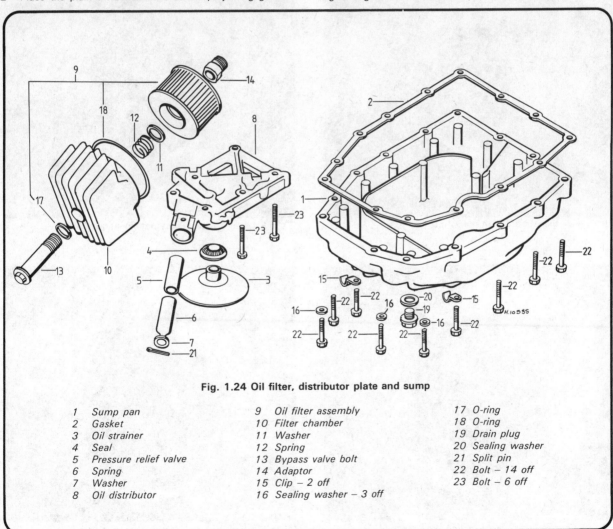

Fig. 1.24 Oil filter, distributor plate and sump

1 Sump pan	9 Oil filter assembly	17 O-ring
2 Gasket	10 Filter chamber	18 O-ring
3 Oil strainer	11 Washer	19 Drain plug
4 Seal	12 Spring	20 Sealing washer
5 Pressure relief valve	13 Bypass valve bolt	21 Split pin
6 Spring	14 Adaptor	22 Bolt – 14 off
7 Washer	15 Clip – 2 off	23 Bolt – 6 off
8 Oil distributor	16 Sealing washer – 3 off	

38.1 Refit sump, noting position of wiring guides

39.1a The primary drive pinion components

39.1b Slide the primary shaft spacer into position

39.1c Place pinion over shaft (note locating dowel)

39.1d Fit outer pinion over projecting shoulder

39.1e Position the large plain washer as shown

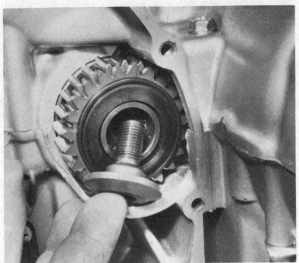
39.1f Fit Belville washer and secure with bolt

39.2a Place the plain thrustwasher over the projecting mainshaft

39.2b Clutch drum bearings should be well lubricated

39.2c Place the clutch outer drum in position

39.3a Fit the splined thrustwasher, followed by ...

39.3b ... the stepped spacer as shown

39.4a Assemble the plain and friction plates on the clutch centre ...

39.4b ... noting the position of the shock absorbing plate

39.5a Use springs and washers to lock clutch ...

39.5b ... then fit tab washer and securing nut

39.6a Fit the remaining clutch springs ...

39.6b ... followed by the release plate and push rod

40 Engine and gearbox reassembly: refitting the starter clutch and automatic timing unit (ATU)

1   Check that the three clutch rollers are in position in the starter clutch body. If any have become displaced, fit the spring and plunger into its recess and hold it in position with a small screwdriver while the roller is fitted. Assemble the clutch body and starter driven gear, and check that the clutch operates normally.

2   Install the starter idler gear and support pin in its recess in the crankcase. Fit the starter clutch and driven gear assembly to the end of the crankshaft, ensuring that the driven gear teeth mesh with those of the idler gear.

3   Offer up the automatic timing unit (ATU), ensuring that the locating pin at the rear of the unit fits into the slot in the shaft end. Hold the crankshaft by fitting a spanner on the large hexagon, then fit and tighten the securing bolt to a torque setting of 3.3 – 3.7 kgf m (24 – 27 lbf ft).

4   Using a new gasket refit the engine left-hand cover, complete with the CDI pickup stator, and tighten its retaining bolts; ensure that the wiring grommet is correctly located at the casing joint. Note that the circular inspection cover should not be refitted until the valve timing has been set as described in Section 44.

40.1a Fit the starter clutch springs and plungers

40.1b Depress plunger whilst roller is inserted

40.1c Check that clutch operates smoothly

40.1d Refit clutch cover plate

40.3 Fit ATU and tighten central securing bolt

41 Engine and gearbox reassembly: refitting the alternator

1   Check that the crankshaft taper and the corresponding tapered bore of the alternator rotor are free from contamination, then place the rotor in position. It will be necessary to prevent crankshaft rotation while the rotor retaining bolt is tightened, using the same method employed during removal. Fit the headed retaining bolt and tighten to 8.0 – 10.0 kgf m (58 – 72 lbf ft).
2   The outer cover carries the stator and brushes, and can now be refitted. Check that the brushes are in a serviceable condition prior to installation, referring to Chapter 6 for details. When the cover is in position, route the output lead through the guides on the crankcase.

42 Engine and gearbox reassembly: replacing the pistons and cylinder block

1   Before replacing the pistons, pad the mouths of the crankcase with rag in order to prevent any displaced component from accidentally dropping into the crankcase.
2   Fit the pistons in their original order with the 'IN' mark on the piston crown towards the rear of the engine.
3   If the gudgeon pins are a tight fit, first warm the pistons to expand the metal. Oil the gudgeon pins and small end bearing surfaces, also the piston bosses, before fitting the pistons.
4   Always use new circlips, **never** the originals. Always check that the circlips are located properly in their grooves in the piston boss. A displaced circlip will cause severe damage to the cylinder bore, and possibly an engine seizure.
5   Place a new cylinder base gasket (dry) over the crankcase mouth. Note that an O-ring is fitted around the four inner rear cylinder head studs; larger holes in the gasket will indicate the correct position. Large diameter O-rings are fitted around each cylinder spigot where it protrudes from the cylinder block casting. Check that the two hollow dowel pins are in position in the crankcase face.
6   Before the cylinder block is fitted, it is essential that the camshaft chain tensioner assembly is in position. If it has been removed in the course of overhauling, reposition it in the camshaft chain tunnel and fit the two domed retaining nuts finger-tight.
7   Carefully lower the cylinder block over the holding studs, using suitable wooden blocks to support it clear of the pistons whilst the camshaft chain is threaded through the tunnel between the bores. This task is best achieved by using a piece of stiff wire to hook the chain through, and pull it up through the

tunnel. The chain must engage with the crankshaft drive sprocket.
8   The cylinder bores have a generous lead in for the pistons at the bottom, and although it is an advantage on an engine such as this to use a piston ring compressor, in the absence of this, it is possible to lead gently the pistons into the bores, working across from one side. Great care has to be taken **not** to put too much pressure on the fitted piston rings. When the pistons have finally engaged, remove the rag padding from the crankcase mouths and lower the cylinder block still further until it seats firmly on the base gasket.
9   Take care to anchor the camshaft chain throughout this operation to save the chain dropping down into the crankcase. The chain should be kept reasonably taut to prevent it from bunching around the crankshaft sprocket. This is particularly important if the crankshaft is turned, as the chain will tend to jam if left to its own devices. Fit and tighten the single cylinder base nut which will be found at the front of the camshaft chain tunnel. If required, there are two Honda service tools which can be usefully employed during this stage of reassembly; a pair of piston support blocks (07958 – 2500001) and a pair of piston ring compressors (07954 – 4220000).

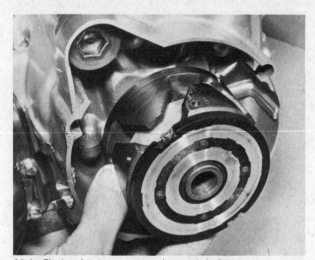

41.1a Fit the alternator rotor to the crankshaft taper ...

41.1b ... and tighten the securing bolt

42.2 Check that 'IN' mark faces towards rear of engine

42.3 Warm the piston to ease the fitting of gudgeon pin

42.4 Ease circlip into position, ensuring that it locates properly

42.5 Fit new base gasket. Note O-rings and dowels

42.8 Check that rings enter bores squarely

43 Engine and gearbox reassembly: refitting the cylinder head

1   If the upper camshaft chain tensioner was removed during the overhaul, it should be refitted in the cylinder head prior to the installation of the latter. Once in position, depress the tensioner blade with the locking bolt slackened to achieve the tensioner's slackest setting, then re-tighten the bolt and locknut. Moving to the lower tensioner assembly in the cylinder barrel, tighten both domed locknuts, then slacken off the lower of the two. Grasp the top of the tensioner blade with pliers and pull it up against the spring pressure. Tighten the lower locknut to retain the setting.

2   Check that the cylinder head and block jointing faces are quite clean and free from traces of old cylinder head gasket. Fit the two dowel pins, and then install a new cylinder head gasket. The cylinder head gasket must be fitted with the wider (5 mm) edges of the individual cylinder periphery seals facing upwards. Lower the cylinder head into position, taking care to ensure that the camshaft chain is fed up through its central aperture. Check that the cylinder head seats squarely and that the dowels have located properly. If necessary, tap the top of the cylinder head using a soft-faced mallet to help seat it.

3   Fit the twelve cylinder head cap nuts, together with a plain washer beneath each one. Care should be exercised when securing the cylinder head, because it can easily become warped if uneven pressure is applied. The nuts should be tightened in a diagonal sequence, working from the centre outwards as shown in the accompanying illustration. Start by tightening the nuts to approximately half the final torque value, then go through the sequence once more to bring the nuts up to full pressure. The cylinder head nut torque setting is 3.6 – 4.0 kgf m (26 – 29 lbf ft).

4   Fit and tighten the two small bolts which pass upwards into the cylinder head from the camshaft chain tunnel flange. Fit the camshaft oil feed pipe in position at the rear of the cylinder block, noting that the pipe is routed between the carburettor adaptors of cylinders 3 and 4. Bear in mind that the well-being of the camshafts and valve gear is entirely dependent upon a reliable oil feed through this pipe, and for this reason the union sealing washers should be renewed as a precautionary measure even if they appear to be in good condition. Note also that the two union bolts differ in that the oil drillings are of different sizes. The bolt with the larger hole **must** be fitted to the upper union. When fitting the bolts, hold the unions with a self-locking wrench so that no strain is placed on the pipe itself as the bolt is tightened.

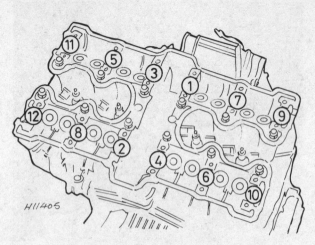

Fig. 1.25 Cylinder head nut tightening sequence

43.4 Tighten the two small bolts at front of cylinder head

44 Engine and gearbox reassembly: refitting the camshafts and setting the valve timing

1   Install the cam followers and adjustment shims in their correct locations, if this has not been done already. The followers and bores should be lubricated with engine oil during installation. Lubricate the camshaft bearing faces with molybdenum disulphide grease.

2   Holding the camshaft chain taut to prevent it from bunching around the crankshaft sprocket, turn the crankshaft by means of the large hexagon on the ATU until the **1,4T** mark appears in the timing window. Line the timing mark up against the index line on the outer cover.

3   Fit the smaller camshaft connecting chain around the smaller section of the exhaust camshaft sprocket ensuring that it is installed to run in its original direction of rotation. The sprocket should now be fitted to the protruding end of the main camshaft chain, taking care not to move the crankshaft. Note that the sprocket has two alignment dots on its left-hand face; these should be arranged horizontally so that they are parallel to the gasket face.

4   Slide the exhaust camshaft (note tachometer drive as means of identification) through the centre of the sprocket, positioning the cam lobes for the No 1 (left-hand) cylinder so that they face horizontally towards the sparking plug. Fit the A and E camshaft bearing caps, securing them by fitting the bolts finger-tight. Note that the arrows on the caps **must** face forwards. At this stage, one of the camshaft sprocket mounting bolt holes should be accessible and in line with the camshaft's threaded hole. Fit a securing bolt, loosely at this stage.

5   Fit the D and the unmarked tachometer drive bearing caps, again with the bolts finger-tight. Note that the D bearing cap has a groove which locates the camshaft.

6   Turn the crankshaft through 360° (one complete revolution) so that the remaining sprocket mounting hole becomes accessible. Fit the second bolt and tighten it to the specified torque setting. Turn the engine through 360° once more, and tighten the first bolt to the same torque figure.

7   Complete the installation of the camshaft bearing caps, not forgetting the locating dowels fitted to each one. Note that each cap has an identification letter which indicates its position (see Section 8 of this Chapter for details). The bearing cap securing bolts should be tightened progressively in a diagonal sequence to a torque setting of 1.2 – 1.6 kgf m (9 – 12 lbf ft).

8   Set the camshaft chain tensioner by slackening the lower of the two cap nuts at the rear of the cylinder block. This will allow the tensioner to find its own setting, after which the nut can be

re-tightened. Check that the **1,4T** mark is still aligned, then re-check that the cam lobes of No 1 cylinder are facing the sparking plug, and that the camshaft sprocket punch marks are parallel to the cylinder head gasket face. If the above checks prove that the exhaust camshaft is correctly timed in relation to the crankshaft, proceed as described below, otherwise re-set the timing until all of the marks align correctly.

9   If the inlet camshaft sprocket was not removed from the camshaft during dismantling or overhaul, loop the connecting cam chain around the sprocket and lower the assembly into position, ensuring that the No 1 cylinder cam lobes face towards the sparking plug and that the punch marks are parallel to the cylinder head gasket face.

10 In cases where the sprocket was removed, fit the chain around the sprocket so that the punch marks lie parallel to the gasket face and in line with those of the exhaust camshaft sprocket. Fit the accessible sprocket bolt finger-tight.

11 Fit the inlet camshaft bearing caps (F, G, K and L) tightening the retaining bolts evenly in a diagonal sequence to 1.2 – 1.6 kgf m (9 – 12 lbf ft). Turn the crankshaft through 360° and fit the remaining camshaft sprocket bolt, tightening it to the specified torque setting, then turn the crankshaft another complete turn and secure the first sprocket bolt to the same torque value.

12 Set the connecting camshaft chain tension by slackening the locking bolt to allow the slack to be taken up. (Note that the chain tension should be re-checked after the engine has been started). Set up the crankshaft timing mark once more, and make a final check on the camshaft timing as described above.

13 Fit the black plastic oil deflector cap on each of the two cylinder head nuts nearest to the camshaft chain tunnel on the inlet side of the cylinder head. Fit the chain tensioner support plate, securing it with its right-hand mounting bolt only. Place the oil feed pipe and chain guide in position. These are retained by the inner bearing cap bolts to the right of the camshaft connecting chain, and by a single bolt on the left, the latter doubling as the means of holding the left-hand side of the support plate.

14 Prime the recesses around each valve with engine oil to provide lubrication when the engine is first started. Check the valve clearances as described in Routine Maintenance, and make any necessary adjustments before proceeding further. Check the cylinder head cover gasket for indentations or other damage. If it is in good condition, it can be re-used. Clean the gasket and gasket face, and apply a smear of RTV sealant in the angled areas formed by the semi-circular end plugs. The cover can now be refitted.

44.2 Align crankshaft as shown for valve timing

44.3a Fit cam chains around exhaust camshaft sprocket

44.3b Alignment dots should lie parallel to gasket face

44.4 Cam lobes must face towards sparking plugs

44.5 Assemble bearing caps as described in text

44.7 Tighten cap bolts to recommended torque figure

44.10 Install the inlet camshaft and secure the sprocket

44.13a Refit chain guide and oil pipe

44.13b Fit oil deflector caps as shown

45 Refitting the engine and gearbox unit into the frame

1   As mentioned during the engine removal sequence, the engine/gearbox unit is unwieldy, requiring at least two, or preferably three, people to coax it back into position. This is even more important during reassembly, as the unit must be offered up at the right angle, and then manoeuvred into position. Care must be taken not to damage the finish on the frame tubes, and it is worthwhile protecting these with rag or masking tape.

2   Needless to say, the unit should be fitted from the right-hand side of the frame, where the removable lower section permits clearance. Again, bear in mind that the engine/gearbox unit is both heavy and awkward to manoeuvre; do not take chances which might result in damage to the machine or the operator. Note that it is important that the oil filter and housing are left **off** until the engine is installed.

3   It is recommended that a trolley jack is used to facilitate installation. This will take the weight of the unit whilst the

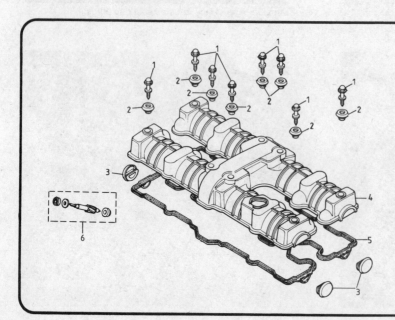

Fig. 1.26 Cylinder head cover

1   Retaining bolt – 8 off
2   Seal – 8 off
3   Plug – 4 off
4   Cover
5   Gasket
6   Tachometer drive gear

mounting bolts are refitted. Always use a wooden block between the crankcase and the jack. Fit the various mounting bolts and plates in position, as shown in the accompanying photographs, but **do not** tighten them until they are all in place. Bear in mind the various footrest/brake pedal arrangements featured on the different models, as described in the removal sequence (Section 6).

4   Once in position, the various bolts and nuts should be tightened to the values specified below.

Torque settings
   8 mm 1.8 – 2.5 kgf m (13 – 18 lbf ft)
   10 mm 3.0 – 4.0 kgf m (22 – 29 lbf ft)
   12 mm 5.5 – 6.5 kgf m (40 – 47 lbf ft)

46 Engine and gearbox unit installation: final assembly and adjustment

1   Place the final drive chain over the rear sprocket, and fit it over the splined end of the layshaft. Refit the bolt, tightening it to the specified torque setting. Lock the rear wheel by applying the rear brake whilst the nut is tightened. Knock over the locking washer to prevent the nut from slackening in use. Note that the chain guide plate should be in position at this stage (see photograph 45.2f). Do not omit to reset the drive chain tension by means of the rear wheel spindle adjusters.

2   If the oil pump was removed, it should be refitted at this stage using a new gasket. Make sure that the pump orifices and gasket faces are absolutely clean, and that the locating dowel is fitted to the lower right-hand mounting bolt. Refit the neutral switch lead clamp to the appropriate oil pump cover bolt (located nearest the centre of the cover's vertical column), then refit the cover and breather hose.

3   Connect the heavy duty starter motor cable to its terminal, and slide the protective rubber boot over the exposed connection. Fit the motor to its recess in the upper crankcase and fit the securing bolts. Reconnect the oil pressure switch lead. Refit the starter motor cover.

4   Refit the alternator and sprocket covers, ensuring that the output leads of the former are correctly routed and re-connected at their connector block. Refit the sparking plug leads, noting that they are numbered 1 to 4, 1 being the left-hand cylinder. Connect the tachometer drive cable at the cylinder head cover and secure the single locking bolt.

5   In the case of the CB900F model, refit the oil cooler assembly to the frame and refit the pipe unions to the underside of the crankcase. Reassemble the exhaust system in the reverse order of that described for removal, using a new sealing ring in each exhaust port.

6   Place the air cleaner casing loosely in position in the frame. The mounting bolts should be left slack until the carburettors have been fitted. Connect the throttle cables to the operating quadrant, and seat the cable outers against their respective stops. Adjust the throttle cable free play to give 2 – 6 mm measured at the flanged end of the throttle grip in relation to the adjacent switch housing. This adjustment can be carried out with the carburettors installed, but access will be easier at this stage.

7   Manoeuvre the carburettor bank into position, displacing the air cleaner casing slightly to obtain sufficient clearance. Persuading the carburettors to engage in the rubber mounting stubs is not easy, and will demand a degree of patient manipulation with the aid of a small screwdriver. The operation is made easier if an assistant is available to deal with one side of the carburettor bank. Tighten the securing clips, then repeat the operation with the air cleaner hose connections. When the carburettor bank is back in place, reposition the air cleaner casing and tighten its mounting screws.

8   Trace and reconnect the CDI pickup leads, plus the alternator leads if these have not been fitted. Reconnect the battery leads and check that the electrical system functions properly. Pay particular attention to the rear brake light switch, which may be in need of adjustment, and to the oil pressure and neutral light switches.

9   Refit the side panels, then slide the fuel tank into position, ensuring that the mounting rubbers at the front of the tank engage correctly. Secure the tank with the single bolt at the rear. Reconnect the petrol feed pipes to the carburettors, and check that the various drain and breather hoses are routed correctly.

10 Assemble the oil filter housing with a new filter element. Check that the O-ring is in sound condition, and renew if necessary. Tighten the filter housing bolt to 2.8 – 3.2 kgf m (20 – 23 lbf ft). Fill the crankcase with 4.5 litres (9.51/7.92 US/Imp pints) of SAE 10W/40 engine oil, noting that the oil level must be checked and topped up after the engine has been started and run for a few minutes.

11 Reconnect the clutch cable to the actuating arm on the outer cover. Set the cable adjuster to give 3/8 – 3/4 in (10 – 20 mm) free play measured at the lever end. Finally, check around the workbench area for any 'left-over' parts; these should be identified and refitted before attempting to start the newly-rebuilt engine.

45.2a Rear lower mounting showing removable frame section

45.2b RH front mounting showing frame section joint

45.2c LH front mounting plates are fitted as shown

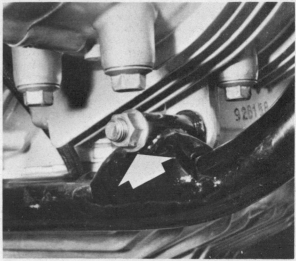
45.2d Through bolt is fitted on underside of unit (arrowed)

45.2e RH upper rear mounting – note earth lead and spacer (arrowed)

45.2f LH upper rear mounting – note chain guide and spacer (arrowed)

46.2a Fit the oil pump using a new gasket

46.2b Fit oil pump cover, noting position of clip

46.4a Refit the gearbox sprocket cover ...

46.4b ... and fit the gearchange pedal or linkage

46.5 Place a new sealing ring in each exhaust port

pressure warning lamp goes out after a few moments running.

3   Before taking the machine on the road, check that the brakes are correctly adjusted, with the required level of hydraulic fluid in the handlebar master cylinder.

4   Make sure the rear chain is correctly tensioned to 5/8 – 1 inch up and down play. Also that the front forks are filled with the correct amount of oil.

5   Check the exterior of the engine for signs of oil leaks or blowing gaskets. Before taking the machine on the road for the first time, check that all nuts and bolts are tight and nothing has been omitted during the reassembling sequence.

48 Taking the rebuilt machine on the road

1   Any rebuilt engine will take time to settle down, even if the parts have been replaced in their original order. For this reason it is highly advisable to treat the machine gently for the first few miles, so that the oil circulates properly and any new parts have a reasonable chance to bed down.

2   Even greater care is needed if the engine has been rebored or if a new crankshaft and main bearings have been fitted. In the case of a rebore the engine will have to be run-in again as if the machine were new. This means much more use of the gearbox and a restraining hand on the throttle until at least 500 miles have been covered. There is not much point in keeping to a set speed limit; the main consideration is to keep a light load on the engine and to gradually work up the performance until the 500 mile mark is reached. As a general guide, it is inadvisable to exceed 4,000 rpm during the first 500 miles and 5,000 rpm for the next 500 miles. These periods are the same as for a rebored engine or one fitted with a new crankshaft. Experience is the best guide since it is easy to tell when the engine is running freely.

3   If at any time the oil feed shows signs of failure, stop the engine immediately and investigate the cause. If the engine is run without oil even for a short period, irreparable engine damage is inevitable.

47 Starting and running the rebuilt engine unit

1   Make sure that all the components are connected correctly. The electrical connectors can only be fitted one way, as the wires are coloured individually. Make sure all the control cables are adjusted correctly. Check that the fuse is in the fuse holder, and try all the light switches and turn on the ignition switch. Close the choke lever to start.

2   Switch on the ignition and start the engine by turning it over a few times with the electric starter, bearing in mind that the fuel has to work through the four carburettors. Once the engine starts, run at a fairly brisk tick-over speed to enable the oil to work up to the camshafts and valves. Check that the oil

49 Fault diagnosis: engine

Symptom	Cause	Remedy
Engine will not start	Defective sparking plugs	Remove the plugs and lay them on the cylinder head. Check whether spark occurs when ignition is on and engine rotated.
	Faulty ignition system	See Chapter 3.

Engine runs unevenly	Ignition or fuel system fault	Check each system independently, as though engine will not start.
	Blowing cylinder head gasket	Leak should be evident from oil leakage where gas escapes.
	Incorrect ignition timing	Check accuracy and reset if necessary.
Lack of power	Fault in fuel system or incorrect ignition timing	Check fuel lines or float chambers for sediment. Reset ignition timing.
Heavy oil consumption	Cylinder block in need of rebore	Check bore wear, rebore and fit oversize pistons if required.

50 Fault diagnosis: clutch

Symptom	Cause	Remedy
Engine speed increases as shown by tachometer but machine does not respond	Clutch slip	Check clutch adjustment for free play, at handlebar lever, check thickness of inserted plates.
Difficulty in engaging gears, gear changes jerky and machine creeps forward when clutch is withdrawn, difficulty in selecting neutral	Clutch drag	Check clutch for too much free-play. Check plates for burrs on tongues or drum for indentations. Dress with file if damage not too great.
Clutch operation stiff	Damaged, trapped or frayed control cable	Check cable and renew if necessary. Make sure cable is lubricated and has no sharp bends.

51 Fault diagnosis: gearbox

Symptom	Cause	Remedy
Difficulty in engaging gears	Selector forks bent Gear clusters not assembled correctly	Replace with new forks. Check gear cluster for arrangement and position of thrust washers.
Machine jumps out of gear	Worn dogs on the ends of gear pinions	Renew worn pinions.
Gear change lever does not return to original position	Broken return spring	Renew spring.

Chapter 2 Fuel system and lubrication

For modifications, and information relating to later models, see Chapter 7

Contents

Specifications

	CB750K(Z)	CB750K LTD	CB750F	CB900F
Fuel tank				
Total capacity	20 litres (4.4/5.3 Imp/ US gallons)	20 litres (4.4/5.3 Imp/ US gallons)	20 litres (4.4/5.3 Imp/ US gallons)	20 litres (4.4/5.3 Imp/ US gallons)
Reserve capacity	5 litres (1.1/1.3 Imp/ US gallons)	5 litres (1.1/1.3 Imp/ US gallons)	4.5 litres (1.0/1.19 Imp/ US gallons)	4.5 litres (1.0.1.19 Imp/ US gallons)
Carburettors				
Make	Keihin	Keihin	Keihin	Keihin
Type	VB42A or VB42C	VB42A	VB42B	VB51A
Primary main jet	68	68	68	68
Secondary main jet	102	102	pre 1980; 98 1980 on; 100	98
Float level	15.5 mm (0.61 in)	15.5 mm (0.61 in)	15.5 mm (0.61 in)	15.5 mm (0.61 in)
Pilot screw setting	VB42A; 1½ turns out VB42C; 1¾ turns out	1½ turns out	pre 1980; 1½ turns out 1980 on; 1⅞ turns out	1½ turns out
Idle speed	1000 ± 100 rpm	1000 ± 100 rpm	1000 ± 100 rpm	1000 ± 100 rpm
Fast idle speed	2000 ± 500 rpm	2000 ± 500 rpm	2000 ± 500 rpm	1000–2500 rpm
Venturi diameter	30 mm (1.18 in)	30 mm (1.18 in)	30 mm (1.18 in)	32 mm (1.26 in)

Lubrication system

Type .... Wet sump, high pressure
Filter .... Pleated paper element, gauze sump strainer
Oil capacity .... 4.5 litres (7.92 Imp pints, 4.7 US quarts)
(3.5 litres, 6.00 Imp pints, 3.7 US quarts at oil changes)
Nominal oil pressure .... 71 psi (5.0 kg cm^2 at 7000 rpm/80°C (176°F)78 psi/5.5 kg
Measured at oil pressure switch   cm^2 at 7000
take off   rpm 80°C
(176°F)

Oil pump

Type	Trochoid			
Delivery rate:				
Engine	41 litres per minute (72 Imp pints per minute, 43.4 US quarts per minute) at 7000 rpm			
Oil cooler	N/A	N/A	N/A	18 litres/32 Imp pints per minute at 7000 rpm
Inner/outer rotor clearance (max)		0.15 mm (0.006 in)		0.20 mm (0.0008 in)
Outer rotor/body clearance (max)		0.35 mm (0.014 in)		
Rotor end clearance (max)		0.10 mm (0.004 in)		

1 General description

The fuel system is comprised of a steel petrol tank from which fuel is fed by gravity to the four Keihin carburettors. The tap has three positions, giving a normal supply of petrol, an emergency reserve position and an off position. A gauze strainer is incorporated in the tap, to trap any foreign matter which might otherwise block the carburettor jets. The petrol tank filler cap incorporates a lock operated by the ignition key.

The four carburettors are interconnected by a linkage to ensure synchronisation. The two throttle cables are connected to a pulley mounted between the two control instruments, the throttles being opened and closed positively. The engine draws air in via a moulded plastic trunking which contains the air cleaner element.

The engine oil is contained in a sump formed at the bottom of the crankcase.

The gearbox is also lubricated from the same source, the whole engine unit being pressure fed by a mechanical oil pump that is driven off the crankshaft. The oil pump intake extends into the sump to pump the oil up to the engine. A screen at the pump inlet point prevents foreign matter from entering the pump before it can damage the mechanism. From the pump the oil passes to the oil filter to be cleaned. If the filter becomes clogged, a safety by-pass valve routes the oil around the filter. It is then routed through a passageway in which an oil pressure switch is mounted, and through an oil hole in the crankcase, from which point it is sent in three different directions. One direction is to the crankshaft main bearings and crankshaft pins. After lubricating the crankshaft parts, the oil is thrown out by centrifugal force and the spray lands on the cylinder walls, the pistons and gudgeon pins to lubricate those parts. The oil eventually drops down from all these points and accumulates in the bottom of the crankcase sump to be recirculated.

The second passageway for oil from the pump takes the form of an oil feed pipe which conveys oil up into the cylinder head. After passing through holes into the camshaft bearings, the oil flows out over the cams and down around the valve tappets to lubricate these areas. The oil returns to the sump via the oil holes at the base of the tappets, and the cam chain tunnel in the centre of the cylinder head and cylinder block.

A third branch of the system feeds oil to the gearbox mainshaft and layshaft, where the gearbox components are positively lubricated.

On CB900F models an oil cooler is fitted to keep the engine oil temperature down to an acceptable level. The oil cooler diverts a proportion of the oil leaving the pump through a pipe to the frame mounted cooler matrix. The cooled oil then returns to the sump, thus lowering the overall engine oil temperature.

2 Petrol tank: removal and replacement

1   The petrol tank fitted to the Honda dohc 4-cylinder models is secured to the frame by means of a short channel that projects from the nose of the tank and engages with a rubber buffer surrounding a pin welded to the frame immediately behind the steering head. This arrangement is duplicated either side of the nose of the tank and the frame. The rear of the tank is secured by a single bolt that passes through a lip welded on to the back of the tank. The tank also has two rubber buffers on which it rests at the rear. A petrol tap is fitted with a reserve pipe that is switched over, when the fuel level falls below that of the main feed pipe.

2   The petrol tank can be removed from the machine without draining the petrol, although the rubber fuel lines to the carburettors will have to be disconnected. The dualseat must be lifted up to expose the mounting bolt at the rear of the tank, then the tank raised at the rear and pulled upwards and backwards to pull off the front rubbers. When replacing the fuel tank, lift at the rear and push down onto the front rubber buffers, then secure the bolt at the rear and reconnect the fuel lines.

3 Petrol tap and filter: removal, dismantling and replacement

1   It is not necessary to drain the petrol tank if it is only half or under half full, as the tank can be laid on its side on a clean cloth or soft material (to protect the enamel), so that the petrol tap is uppermost. The petrol pipe should be removed before unscrewing the petrol tap. The tap is released by slackening the gland nut which secures it to its mounting stub on the underside of the tank.

2   The gauze strainer can be removed for cleaning by pulling it upwards, clear of the main pickup pipe. No further maintenance is feasible, as parts are not available for the tap. In the event of malfunction it must be renewed. Do not forget that fuel leakage can be dangerous, and may be illegal in some parts of the world.

3.1 The fuel tap is secured to tank by a gland nut

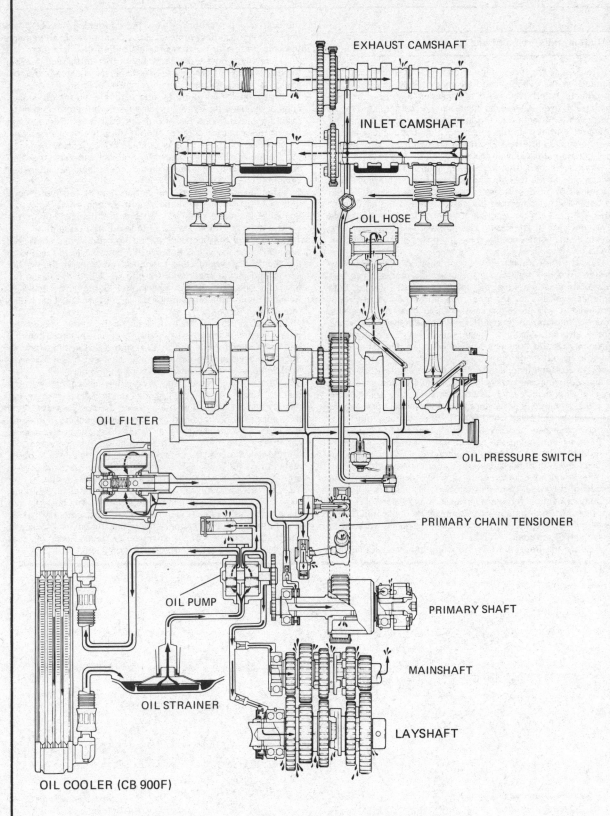

EXHAUST CAMSHAFT

INLET CAMSHAFT

OIL HOSE

OIL FILTER

OIL PRESSURE SWITCH

PRIMARY CHAIN TENSIONER

OIL PUMP

PRIMARY SHAFT

MAINSHAFT

OIL STRAINER

LAYSHAFT

OIL COOLER (CB 900F)

Fig. 2.1 Lubrication system

Note that oil cooler is fitted to CB900F models only.

4   Carburettors: removal and separation

1   The four constant depression (CD) carburettors are mounted on the cylinder head as an assembly, the individual instruments being connected by linkages and by mounting brackets. If overhaul proves necessary, a certain amount of work, such as attention to the float assembly and jets, can be carried out with the instruments in position, but because access is very restricted it is best to remove the complete bank so that work can be carried out on the bench.

2   Start by ensuring that the fuel tap is turned off, and remove the fuel pipe at the fuel tap end. Lift the seat and remove both side panels to provide access to the air cleaner casing. Slacken the hose clips which secure the carburettors to the air cleaner hoses, then release the single bolt which retains the top of the air cleaner casing. The casing can now be pulled rearwards to provide a small clearance between the hoses and the carburettors.

3   Slacken the carburettor mounting clips. With the aid of an assistant, where possible, pull the assembly rearwards to free the mounting stubs from the inlet adaptors. There is little clearance available for this operation, and a degree of careful manoeuvring will prove necessary. Once clear of the stubs, withdraw the carburettors sufficiently to permit access to the throttle and choke cables. These should be disconnected by slackening off the adjusters and locknuts to allow the inner cables to be released.

4   Further dismantling is dependent on the degree of overhaul anticipated; generally speaking, it is possible to attend to most areas of the instruments without separating them. Should complete dismantling prove necessary, bear in mind that a lot of work will be involved. Additionally, the manufacturer recommends that new choke valves, shafts and screws are fitted.

5   Start by unhooking the end of the relief spring, which is located between the No 3 and 4 carburettors. The spring is the lighter of the two fitted concentric to the choke shaft. Holding the synchronising screws with a screwdriver, slacken the locknuts of each one. Turn each screw slowly inwards, counting the number of turns required to seat it. Make a careful note of each figure so that the correct setting can be duplicated during reassembly. Unscrew each of the synchronising screws to release spring tension.

6   Release the front and rear mounting brackets, noting that the screws will probably be very tight, requiring the use of an impact driver to effect removal. Take great care not to damage the carburettors or the screw heads during this operation.

7   Carefully separate the assembly at the centre joint, leaving the carburettors as two pairs. Take care not to damage the fuel and air connecting pipes or the mechanical linkage.

8   Open the choke butterfly and carefully file off the staked ends of the securing screws, holding the carburettors so that the resulting metal filings fall out of the main bore. Remove and discard the securing screws, and lift the butterfly plates away. Release the fuel inlet T-piece retainer, which is held by one of the vacuum chamber screws. Carefully separate the two carburettors, with the same caution described earlier in this section.

9   The linkage between the carburettors is best left alone unless it is absolutely essential that it is removed. If removal is unavoidable, make detailed sketches of the position of the various springs and levers as a guide during reassembly.

10   Disengage the fine relief spring from the choke shaft end, and withdraw the shaft. Remove the split pin from the end of the accelerator pump return spring rod, and remove the washer spring and spring seat. Unscrew the pivot bolt to release the fast idle arm, spring and accelerator pump lever and rod. Release the throttle quadrant by driving out the small pin which retains it to its shaft.

11   Reassembly is a direct reversal of the above sequences, noting the following points. The carburettors should be built up into pairs, and then joined at the centre. Ensure that new O-rings are used on the various connecting pipes and T-pieces, lubricating each one with a smear of engine oil.

12   When fitting the throttle connecting links together, note that the forked arm must fit between the two plain washers, not directly against the spring, the larger of the two washers facing the spring. Fit the front and rear mounting brackets, tightening the screws progressively in a diagonal sequence to preserve carburettor alignment. Fit the thrust springs between the throttle link of each pair of carburettors. Fit the choke butterfly plates using the new screws with tabwashers. Lock the screws after tightening by bending up the locking tabs.

13   Set the synchronising screws to their original positions, then check that the distance between the throttle butterfly and the pilot bypass orifice is identical for each instrument. When all four butterflies are synchronised, tighten the adjuster locknuts. Check the operation of the linkages and levers, ensuring that they operate smoothly and do not bind or jam.

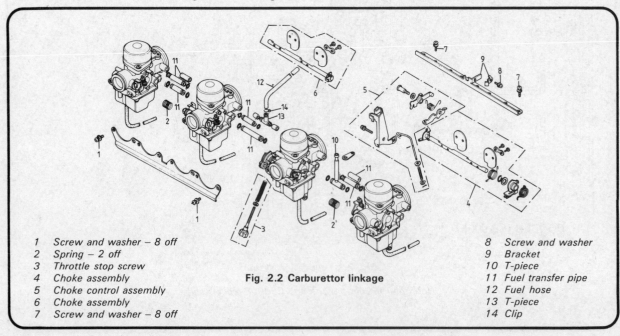

Fig. 2.2 Carburettor linkage

1   Screw and washer – 8 off
2   Spring – 2 off
3   Throttle stop screw
4   Choke assembly
5   Choke control assembly
6   Choke assembly
7   Screw and washer – 8 off
8   Screw and washer
9   Bracket
10   T-piece
11   Fuel transfer pipe
12   Fuel hose
13   T-piece
14   Clip

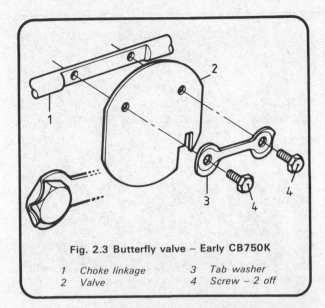

Fig. 2.3 Butterfly valve – Early CB750K

1 Choke linkage
2 Valve
3 Tab washer
4 Screw – 2 off

4.3a Slacken screw (A) to release cable, then unhook from lever (B)

4.3b Throttle cables can be released after slackening adjusters

4.5a Release the choke relief spring (arrowed)

4.5b Note screw setting, then back off to release spring tension (arrowed)

4.7 Separate carburettors at centre joint

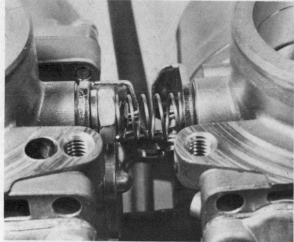

4.8a This small spring will drop free as carburettor pairs are separated

4.8b Remove choke plate from one carburettor

4.8c Draw instruments apart as shown

4.8d Fuel pipe is held by vacuum cylinder screw

4.10 Central carburettor linkages – US models have throttle pump

5　Carburettors: dismantling, examination and renovation

1　As mentioned previously, most of the normal overhaul jobs may be carried out with the instruments joined as a bank of four, thus avoiding a considerable amount of dismantling work. Note that where attention to the connecting linkages is required, it will be necessary to separate the instruments as described in Section 4.

2　Before any dismantling work takes place, drain out any residual fuel and clean the outside of the instruments thoroughly. It is essential that no debris finds its way inside the carburettors.

3　It is suggested that each carburettor is dismantled and reassembled separately, to avoid mixing up the components. The carburettors are handed and therefore components should not be interchanged.

4　Invert one carburettor and remove the three float chamber screws. Lift the float chamber from position and note the chamber sealing ring. This need not be disturbed unless it is damaged. The two floats, which are interconnected, can be lifted away after displacing the pivot pin. The float needle is attached to the float tang by a small clip. Detach the clip from the tang and store the needle in a safe place.

5   Prise out the rubber blanking plug to expose the slow jet for examination. Note that it is pressed into the carburettor body and cannot be renewed, and for the same reason cleaning can only be carried out by using compressed air in the jet passage. When unscrewing any jet, a close fitting screwdriver must be used to prevent damage to the slot in the soft jet material. Hold the secondary main jet holder with a small spanner and unscrew the secondary jet. The holder may then be unscrewed to release the needle jet which is a push fit and projects into the carburettor venturi. Unscrew the main jet from the final housing and then unscrew the main nozzle from the same housing.

6   Unscrew the retaining screws which hold the carburettor cap (piston chamber) and pull the cap from position. Remove the helical spring and the nylon sealing ring. Pull the piston up and out of its slider. The piston needle can be removed by unscrewing the plug in the top of the piston. The needle will drop out. The main air jet and secondary air jet are hidden below a plate, which is retained in the upper chamber by a single cross head screw. Remove the screw and plate. The two slow air jets are similarly positioned opposite the main air jets, but are not closed by a plate. None of these jets can be removed. They must be cleaned in place.

7   A diaphragm air cut-off valve is fitted to each carburettor to richen automatically the mixture on over-run, thus preventing backfiring in the exhaust system. The valve is located on the side of the main body, and thus will require separation of the instruments if attention is required. The cut-off valve is enclosed by a cover held on the outside of the carburettor body by two screws. Unscrew the screws holding the cover in place against the pressure of the diaphragm spring, and then lift the cover away. Remove the spring, and carefully lift out the diaphragm.

8   Check the condition of the floats. If they are damaged in any way, they should be renewed. The float needle and needle seating will wear after lengthy service and should be inspected carefully. Wear usually takes the form of a ridge or groove, which will cause the float needle to seat imperfectly. If damage to the seat has occurred the carburettor body must be renewed because the seat is not supplied as a separate item.

9   After considerable service the piston needle and the needle jet in which it slides will wear, resulting in an increase in petrol consumption. Wear is caused by the passage of petrol and the two components rubbing together. It is advisable to renew the jet periodically in conjunction with the piston needle.

10 Inspect the cut-off valve diaphragm for signs of perishing or perforation. Damage will be easily seen.

11 Before the carburettors are reassembled, using the reversed dismantling procedure, each should be cleaned out thoroughly using compressed air. Avoid using a piece of rag since there is always risk of particles of lint obstructing the internal passage-ways or the jet orifices.

12 Never use a piece of wire or any pointed metal object to clear a blocked jet. It is only too easy to enlarge the jet under these circumstances and increase the rate of petrol consumption. If the compressed air is not available, a blast of air from a tyre pump will usually suffice.

13 Do not use excessive force when reassembling a carburettor because it is easy to shear a jet or some of the smaller screws. Furthermore, the carburettors are cast in a zinc-based alloy which itself does not have a high tensile strength. Take particular care when replacing the throttle valves to ensure the needles align with the jet seats.

14 Do not remove either the throttle stop screw or the pilot jet screw without first making note of their exact positions. Failure to observe this precaution will make it necessary to re-synchronise the carburettors on reassembly.

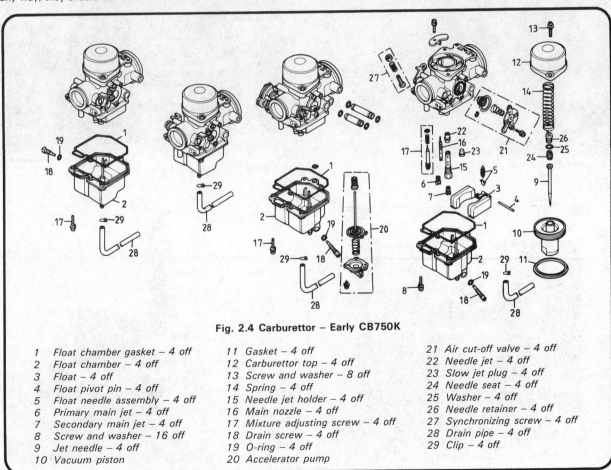

Fig. 2.4 Carburettor – Early CB750K

1 Float chamber gasket – 4 off	*11 Gasket – 4 off*	*21 Air cut-off valve – 4 off*
2 Float chamber – 4 off	*12 Carburettor top – 4 off*	*22 Needle jet – 4 off*
3 Float – 4 off	*13 Screw and washer – 8 off*	*23 Slow jet plug – 4 off*
4 Float pivot pin – 4 off	*14 Spring – 4 off*	*24 Needle seat – 4 off*
5 Float needle assembly – 4 off	*15 Needle jet holder – 4 off*	*25 Washer – 4 off*
6 Primary main jet – 4 off	*16 Main nozzle – 4 off*	*26 Needle retainer – 4 off*
7 Secondary main jet – 4 off	*17 Mixture adjusting screw – 4 off*	*27 Synchronizing screw – 4 off*
8 Screw and washer – 16 off	*18 Drain screw – 4 off*	*28 Drain pipe – 4 off*
9 Jet needle – 4 off	*19 O-ring – 4 off*	*29 Clip – 4 off*
10 Vacuum piston	*20 Accelerator pump*	

5.4a Release screws and lift float bowl away

5.4b Displace the float pivot pin ...

5.4c ... and remove float together with needle

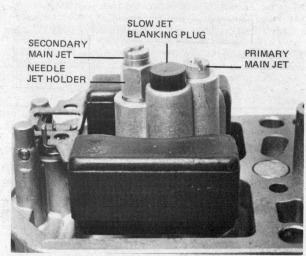

SLOW JET
BLANKING PLUG

SECONDARY
MAIN JET

NEEDLE
JET HOLDER

PRIMARY
MAIN JET

5.5a Carburettor jet location

5.5b Slow jet cannot be removed from carburettor (arrowed)

5.5c Primary main jet can be removed to reveal main nozzle

5.5d Secondary main jet is removed together with needle jet holder

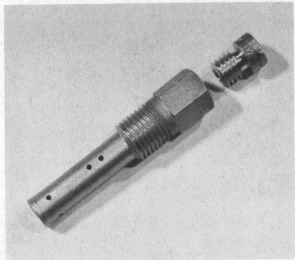
5.5e Secondary main jet can be removed for inspection

5.5f Jet sizes are marked – do not interchange them!

5.6a Remove piston chamber and spring

5.6b Remove piston assembly

5.6c Prise out rubber blanking plug

5.6d Needle is retained by large grub screw

5.6e Damper ring should be renewed if compressed or worn

5.7a Air cut-off valve should be checked for damage

5.7b Remove diaphragm and check for splits

5.7c The air cut-off valve components

5.7d Do not omit this small O-ring during assembly

5.8a Check float needle tip – renew if marked or ridged

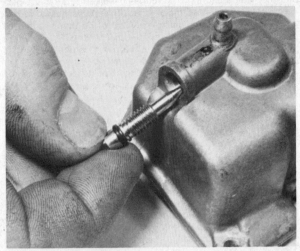

5.8b Float bowl has tapered drain screw

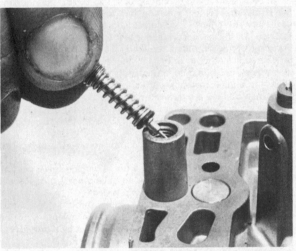

5.11a Pilot screw can be removed for cleaning and inspection

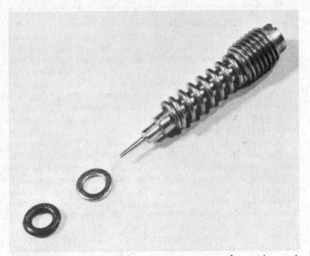

5.11b Pilot screw assembly – note arrangement for washer and O-ring

5.13 Remember fuel pipe guides during reassembly

6 Fast idle mechanism: adjustment

1   The fast idle mechanism takes the form of a pivoted arm mounted between the throttle and choke shafts on the No 2 carburettor. As the choke is operated, a cam on the choke shaft end bears upon the fast idle arm. Movement is transferred to the throttle stop, opening the throttle by a small amount to raise the cold start idle speed to 2000 ± 500 rpm.

2   When the system is off, ie with the throttle valves closed and the choke valves fully open, there should be a small clearance between the fast idle arm and the throttle stop. Measure the gap, which must be within the range 0.7 – 1.0 mm (0.03 – 0.04 in). Any adjustment can be made by carefully bending the forked end of the arm.

7 Accelerator pump: examination and adjustment

1   The US version of the CB750K is equipped with an accelerator pump mounted on the underside of the No 2 carburettor. Its purpose is to richen the mixture during accelera-tion, thus allowing the carburettor to be jetted for a weaker overall mixture to meet EPA emission requirements.

2   The pump is operated by a sprung rod connected to the

throttle cable quadrant. The rod is connected to a lever which terminates in a small metal tang. This depresses a rod to actuate the diaphragm pump. Fuel from the pump is fed by injection nozzles in each of the four instruments.

3   In the event of pump failure, the pump can be dismantled and checked in the same way as already described for the air cut-off valve (see Section 5.7). Little can go wrong with the pump apart from a cracked or perished diaphragm, but if the pump system is completely dry it may require priming to expel air.

4   The pump output is set up during manufacture but should be re-checked after the carburettors have been overhauled or the operating linkage has been disturbed. With the throttle valve closed, check the clearance between the accelerator pump rod and the operating tang. This should be 0.00 – 0.04 mm (0.00 – 0.016 in). Any necessary adjustment may be made by careful bending of the operating tang.

5   At the other end of the arm, the pump stroke is limited by a second tang which stops against the projecting lug on the carburettor body. The specified gap here is 3.1 – 3.3 mm (0.12 – 0.13 in). Once again, adjustment can be made by bending the tang.

8   Carburettors: synchronisation

1   For the best possible performance it is imperative that the carburettors are working in perfect harmony with each other. At any given throttle opening if the carburettors are not synchronised, not only will one cylinder be doing less work but it will also in effect have to be carried by the other cylinders. This effect will reduce the performance considerably. In the case of multi-cylinder engines especially, poor carburettor synchronisation will make the engine feel 'rough' and can induce backlash in the valve train or primary drive, thus giving the illusion of mechanical wear in these areas.

2   Synchronisation is carried out with the aid of vacuum gauges connected to the engine side of the carburettors. The gauges can take the form of mechanical clock-type instruments or a series of glass or plastic tubes, each containing a column of mercury. The latter type is often referred to as a mercury manometer. A suitable vacuum gauge set may be purchased from a Honda Service Agent, or from one of the many suppliers who advertise regularly in the motocyle press.

3   Bear in mind that this equipment is not cheap, and unless the machine is regarded as a long-term purchase, or it is envisaged that similar multi-cylinder motorcycles are likely to follow it, it may be better to allow a Honda dealer to carry out the work. The cost can be reduced considerably if a vacuum gauge set is purchased jointly by a number of owners. As it will be used fairly infrequently this is probably a sound approach.

4   If the vacuum gauge set is available, proceed as follows. Remove the dualseat and petrol tank so that access can be gained to the carburettors. Using a suitable length of feed pipe, reconnect the petrol tank with the carburettors, so that the petrol flow can be maintained. The petrol tank must be placed above the level of the carburettors. Connect the vacuum gauges to the engine.

5   Start the engine and allow it to run until normal working temperature has been reached. This should take 10-15 minutes. Set the throttle so that an engine speed of 1000 ± 100 rpm is maintained. If the readings on the vacuum gauges vary by more than 60 mm Hg (2.4 in Hg) it will be necessary to adjust the synchronising screws to bring the carburettors within limits. Note that if the readings on the gauges fluctuate wildly, it is likely that the gauges require heavier damping. Refer to the gauge manufacturer's instructions on setting up procedures.

6   The No 2 carburettor (second from left) is regarded as the base instrument; that is, it is non-adjustable and the remaining three carburettors must be adjusted to it. Honda produce a special combined screwdriver and socket spanner for dealing with the synchronising screws (Part number 07908-4220100). Its use makes the procedure easier, but it is not essential.

Slacken the locknut of the adjuster concerned, then turn the latter, noting the effect on the gauge reading. When the reading is as close as possible to that of the No 2 carburettor, hold the adjuster screw and retighten the locknut. Repeat the procedure on the remaining carburettors.

9   Carburettors: idle adjustment

CB750 models

1   The engine idle speed should be checked and reset after the synchronizing operation has been carried out as described in the previous section. Before adjusting the carburettors a check should be made to ensure that the following settings are correct: contact breaker gap, ignition timing, valve clearance, sparking plug gaps, crankcase oil level. It is also important that the engine is at normal running temperature.

2   The pilot screws are fitted vertically in each carburettor, adjacent to the float bowl. These are set during assembly and should not be touched unless the carburettors have been overhauled. The master throttle stop screw is located at the centre of the carburettor bank and terminates in a knurled plastic knob. This should be set to give an idle speed of 1000 ± 100 rpm with the engine at normal temperature.

3   If consistent idling cannot be obtained by this method, and the carburettor synchronisation has been checked, it will be necessary to check and adjust the individual pilot screw settings. To do this accurately, a test tachometer calibrated in 50 rpm increments will be required. The machine's tachometer is not sufficiently accurate for the test. Connect the test tachometer according to the manufacturer's instructions and allow the engine to reach normal operating temperature. The adjustment stages are detailed below.

4   Set each pilot screw to its nominal setting (CB750 with VB42A carburettor: $1\frac{1}{2}$ turns out, CB750 with VB42C carburettor: $1\frac{3}{4}$ turns out).

5   Set the master throttle stop control to give the prescribed idle speed of 1000 ± 100 rpm.

6   Unscrew each pilot screw by $\frac{1}{2}$ turn. If engine speed rises by 50 rpm or more, turn the screws out by another $\frac{1}{2}$ turn. Repeat until engine speed drops by 50 rpm or less.

7   Reset the idle speed using the master throttle stop control.

8   Turn the No. 1 carburettor pilot screw inwards until the engine speed drops by 50 rpm, then back it out by $\frac{3}{8}$ turn for the VB42A and by $\frac{3}{4}$ turn in the case of the VB42C.

9   Correct the idle speed once more.

10 Carry out the sequence detailed in paragraphs 8 and 9 for the remaining carburettors.

11 The above sequence is not detailed for CB900F models, and in this case it must suffice to set the pilot screws to their nominal $1\frac{1}{2}$ turns out. In most instances it will be sufficient to set the pilot screws on the CB750 models in the same way, assuming that a tachometer is not available. It follows that this simplified procedure will not produce such accurate results as those obtained by the methods described earlier.

10 Carburettor settings

1   Some of the carburettor settings, such as the sizes of the needle jets, main jets and needle positions, etc, are pre-determined by the manufacturer. Under normal circumstances it is unlikely that these settings will require modification, even though there is provision made. If a change appears necessary, it can often be attributed to a developing engine fault.

2   Always err slightly on the side of a rich mixture, since a weak mixture will cause the engine to overheat. Reference to Chapter 3 will show how the condition of the sparking plugs can be interpreted with some experience as a reliable guide to carburettor mixture strength. Flat spots in the carburation can usually be traced to a defective timing advancer. If the advancer action is suspect, it can be detected by checking the ignition timing with a stroboscope.

11 Carburettors: adjusting float level

1   If problems are encountered with fuel overflowing from the float chambers, which cannot be traced to the float/needle assembly or if consistent fuel starvation is encountered, the fault will probably lie in maladjustment of the float level. It will be necessary to remove the float chamber bowl from each carburettor to check the float level.

2   If the float level is correct the distance between the upper-most edge of the floats and the flange of the mixing chamber body will be 15.5 mm (0.61 in).

3   Adjustments are made by bending the float assembly tang (tongue) which engages with the float tip, in the direction required (see accompanying diagram).

12 Exhaust system

1   Unlike a two-stroke, the exhaust system does not require such frequent attention because the exhaust gases are usually of a less oily nature.

2   Do not run the machine with the exhaust baffles removed, or with a quite different type of silencer fitted. The standard production silencers have been designed to give the best possible performance, whilst subduing the exhaust note to an acceptable level. Although a modified exhaust system, or one without baffles, may give the illusion of greater speed as a result of the changed exhaust note, the chances are that performance will have suffered accordingly.

13 Air filter: removing and cleaning the element

1   The air filter is housed in the plastic trunking to the rear of the carburettors. The element can be removed for cleaning after releasing the left-hand side panel to reveal the access plate. This is secured by two screws. The element itself is retained in the casing by a leaf spring arrangement. The filter is released by pulling the spring out of the casing.

2   The filter is of the pleated paper type and is supported by a metal framework. It can be cleaned by tapping it to dislodge any loose dust, and then blowing compressed air through from the inside. Check the paper surface for tears or contamination, renewing the element if it is excessively dirty or damaged.

3   On no account run the engine without the air cleaner attached, or with the element missing. The jetting of the carburettors takes into account the presence of the air cleaner and engine performance will be seriously affected if this balance is upset.

4   To replace the element, reverse the dismantling procedure. Give a visual check to ensure that the inlet hoses are correctly located and not kinked, split or otherwise damaged. Check that the air cleaner case is free from splits or cracks.

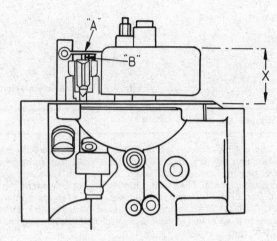

Fig. 2.5 Checking float level

A   *Float tongue*
B   *Float valve*

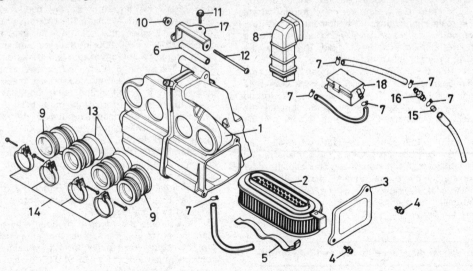

Fig. 2.6 Air cleaner – Early CB750K

1 *Air cleaner case*	7 *Clip – 5 off*	13 *Inner air inlet hose – 2 off*
2 *Element*	8 *Duct*	14 *Hose clamp – 4 off*
3 *End cap*	9 *Outer air inlet hose – 2 off*	15 *Filter*
4 *Screw and washer – 2 off*	10 *Nut*	16 *Union*
5 *Spring*	11 *Bolt*	17 *Case mounting bracket*
6 *Distance spacer*	12 *Screw*	

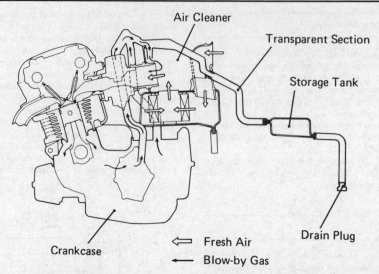

Fig. 2.7 Crankcase emission control system – USA models only

14 Engine lubrication

1   The engine shares a common lubrication system with the gearbox and primary transmission. Oil is picked up through a gauze strainer from the wet sump, and is drawn through the trochoidal oil pump. The pump is mounted externally on the left-hand side of the unit and is driven via gears from the primary shaft. Oil leaving the pump is controlled by a pressure relief valve which maintains a nominal pressure of 71 psi (5.0 kg cm^2) at 7000 rpm.

2   The oil splits into two routes at this juncture, one of which directs the lubricant to the gearbox shafts where the bearings and pinions receive direct pressure lubrication. The oil enters the shaft centres from the left-hand end via a double seal arrangement, and exits via holes in the shafts to the pinion centres.

3   The main engine oil feed is conducted to the engine oil filter at the front of the crankcase where any small impurities are removed by the resin-impregnated filter element. The central filter bolt incorporates a bypass valve which operates in the event that the filter element becomes choked. This allows the oil to continue to circulate, albeit unfiltered.

4   After leaving the filter, the main oil supply is fed to the crankshaft main bearings via an oil gallery in the crankcase casting. Drillings in the crankshaft route the oil to the big-end bearings. The emerging oil splashes serve to lubricate the pistons, cylinder walls and small-end bearing before running back to the sump.

5   A take-off point in the gallery conducts oil through an external pipe to the cylinder head, where the camshafts and valve gear are lubricated. Secondary feeds between the filter and oil gallery supply oil to the primary shaft and to the hydraulic primary chain tensioner. An oil pressure switch is mounted in the gallery to operate a warning light in the event of an oil system failure.

15 Oil pump: dismantling, examination and reassembly

1   The oil pump can be removed with the engine unit in or out of the frame. It is mounted on the left-hand side of the crankcase, and can be reached after the left-hand rear cover and the oil pump cover have been removed. It may prove necessary to remove the footrests and gearchange pedal where a rear-set linkage is not employed.

2   The pump is mounted within a cover, this being retained by seven bolts. Note the position of the neutral switch lead clip for reference during reassembly. The pump can be lifted away after its four mounting bolts have been released. Note that the two countersunk screws should not be touched at this stage.

3   Clean the pump body carefully to avoid any dirt entering the interior. Release the two securing screws and lift away the base plate. If it proves reluctant to move, it is likely that it is being held by the hollow dowel pin located in the lower right-hand mounting hole. This can be tapped out to free the base plate.

4   Displace the inner and outer rotor and remove the drive spindle and gear after withdrawing the small driving pin.

5   Wash all the pump components with petrol and allow them to dry before carrying out an examination. Before partially reassembling the pump for various measurements to be carried out, check the casing for breakage or fracture, or scoring on the inside perimeter.

6   Reassemble the pump rotors and measure the clearance between the outer rotor and the pump body, using a feeler gauge. If the measurement exceeds the service limit of 0.35 mm (0.014 in) the rotor or the body must be renewed, whichever is worn. Measure the clearance between the outer rotor and the inner rotor, using a feeler gauge. If the clearance exceeds 0.15 mm (0.006 in) the rotors must be renewed as a set. It should be noted that one face of the outer rotor is punch marked. The punch mark should face away from the main pump casing during measurements and on reassembly. With the pump rotors installed in the pump body lay a straight edge across the mating surface of the pump body. Again with a feeler gauge measure the clearance between the rotor faces and the straight edge. If the clearance exceeds 0.10 mm (0.004 in) the rotors should be renewed as a set.

7   Examine the rotors and the pump body for signs of scoring, chipping or other surface damage which will occur if metallic particles find their way into the oil pump assembly. Renewal of the affected parts is the only remedy under these circumstances, bearing in mind that the rotors must always be renewed as a matched pair.

8   Reassemble the pump components by reversing the dismantling procedure. Remember that the punch marked face of the rotor must face away from the main pump body. The component parts must be ABSOLUTELY clean or damage to the pump will result. Replace the rotors and lubricate them thoroughly, before refitting the cover. Refit the hollow dowel before replacing the cover plate and tightening the screws.

9   Place a new gasket in position and offer the pump up so that the driven gear engages with the drive gear in the casing. Insert the hollow dowel pin. Fit and tighten evenly the screws.

10 Install the oil pump cover, securing the neutral switch lead by means of the clamp. Assemble the remaining components (where relevant) by reversing the dismantling procedure.

15.2 A: pump retaining bolts. B: pump cover screws

15.3 Release cover screws and separate pump body and cover

15.4 Remove rotors and check pump for wear and damage

15.6a Measure outer rotor to pump body clearance

15.6b Check clearance between inner and outer rotors

15.6c Check end float using straight edge across faces

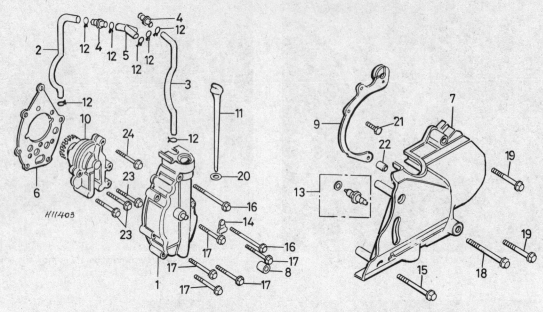

H11403

Fig. 2.8 Oil pump assembly

1 Oil pump cover	9 Guard	17 Bolt – 2 off
2 Breather hose	10 Oil pump	18 Bolt
3 Breather hose	11 Filler cap/dipstick	19 Bolt – 2 off
4 Union – 2 off	12 Clip – 7 off	20 O-ring
5 Y-piece	13 Neutral switch	21 Bolt
6 Gasket	14 Clip	22 Dowel pin – 2 off
7 Outer cover	15 Bolt	23 Bolt – 3 off
8 Rubber stop	16 Bolt – 2 off	24 Bolt

16 Oil strainer and pressure relief valves: location and cleaning

1   The oil strainer assembly and pressure relief valves are housed within the sump, and can be checked after the latter has been detached. Although not detailed in the routine maintenance schedule, it is worth cleaning the sump area occasionally because the nature of the sediment which collects in the sump will give an indication of the overall condition of the machine. If there is any suspicion of oil pressure failure, the relief valves should be checked first.

2   Remove the fourteen sump bolts, having first drained off the engine oil. Lift the sump away. The strainer and main relief valve form part of a unit which is bolted to the underside of the crankcase, and this can be detached if desired. The second relief valve is located forward of the main assembly, and is responsible for maintaining the correct pressure within the hydraulic chain tensioner. Both relief valves can be dealt with in a similar manner.

3   Detach the gauze strainer and flush it clean with petrol or a low flash-point solvent where available. Take note of any metallic particles found, because these can often be linked to unusual noises that may have been noticed. Bronze-coloured particles will indicate that a bush is wearing or breaking up, whilst greyish particles may be the result of wear in the main or big-end bearings, or in the casings. Obviously, any large pieces of metal will warrant further investigation.

4   The relief valves can be dismantled by releasing the split pin whilst depressing the spring with a suitable screwdriver. Release spring tension gradually to avoid the risk of damage or injury. Inspect the bore and plunger for wear, damage or contamination, cleaning or renewing the necessary parts as required. During reassembly, lubricate the plunger with oil, having made sure that the valve components are perfectly clean. When fitting the sump, use a new gasket where

necessary. Tighten the sump bolts in a diagonal sequence to 1.0 – 1.4 kgf m (7 – 10 lbf ft).

17 Oil filter: renewing the element

1   The oil filter is contained within a semi-isolated chamber at the front of the crankcase. Access to the element is made by unscrewing the filter cover centre bolt, which will bring with it the cover and also the element. Before removing the cover, place a receptacle beneath the engine, to catch the engine oil contained in the filter chamber.

2   When renewing the filter element it is wise to renew the filter cover O-ring at the same time. This will obviate the possibility of any oil leaks. Do not overtighten the centre bolt on replacement; the correct torque setting is 2.8 – 3.2 kgf m (20 – 23 lbf ft).

3   The filter by-pass valve, comprising a plunger and spring, is situated in the bore of the filter cover centre bolt. It is recommended that the by-pass valve be checked for free movement during every filter change. The spring and plunger are retained by a pin across the centre bolt. Knocking the pin out will allow the spring and plunger to be removed for cleaning.

4   Never run the engine without the filter element or increase the period between the recommended oil changes or oil filter changes. Engine oil should be changed every 4000 miles, as should the filter element. Use only the recommended viscosity of oil.

18 Oil pressure warning lamp

1   An oil pressure warning lamp is incorporated in the lubrication system to give immediate warning of excessively low oil pressure.

2   The oil pressure switch is screwed into the crankcase to the rear of the cylinder block. The switch is interconnected with a warning light on the lighting panel on the handlebars. The light should be on whenever the ignition is on but will usually go out at about 1500 rpm.

3   If the oil warning lamp comes on whilst the machine is being ridden, the engine should be switched off immediately, otherwise there is a risk of severe engine damage due to lubrication failure. The fault must be located and rectified before the engine is re-started and run, even for a brief moment. Machines fitted with plain shell bearings rely on high oil pressure to maintain a thin oil film between the bearing surfaces. Failure of the oil pressure will cause the working surfaces to come into direct contact, causing overheating and eventual seizure.

4   If low oil pressure is experienced, it can be checked by connecting a suitable gauge to the switch take-off. At normal operating temperatures, a reading of 5.0 kg cm^2 (71 psi) at 7000 rpm should be indicated. Any substantial drop in pressure should be investigated promptly if engine damage is to be avoided.

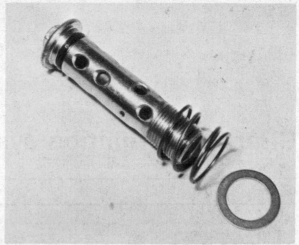

17.3 Check filter bypass valve/bolt for free movement

19 Fault diagnosis: fuel system and lubrication

Symptom	Cause	Remedy
Engine gradually fades and stops	Fuel starvation Sediment in filter bowl or float chamber	Check vent hole in filler cap Dismantle and clean
Engine runs badly. Black smoke from exhausts	Carburettor flooding	Dismantle and clean carburettor. Check for punctured float or sticking float needle
Engine lacks response and overheats	Weak mixture Air cleaner disconnected or hose split Modified silencer has upset carburation	Check for partial block in carburettors Reconnect or renew hose Replace with original design
Oil pressure warning light comes on	Lubrication system failure	Stop engine immediately. Trace and rectify fault before re-starting
Engine gets noisy	Failure to change engine oil when recommended	Drain off old oil and refill with new oil of correct grade. Renew oil filter element

Chapter 3 Ignition system

For modifications, and information relating to later models, see Chapter 7

Contents

Specifications

Ignition system

	750 models	900 model
Type	Capacitor discharge ignition (CDI)	

Ignition timing – UK models

	750 models	900 model
Retarded	10° BTDC @ idle	10° BTDC @ idle
Advanced	30° BTDC @ 6000 rpm	28.5° BTDC @ 3100 rpm

Ignition timing – US models

Retarded	10° BTDC @ 1000 rpm idle
Advanced	40° BTDC @ 6000 rpm
	36° BTDC @ 7400 rpm

Sparking plugs

	750 models (UK)	750 models (US)	900 model
Make	NGK or ND	NGK or ND	NGK or ND
Type	DR8ES-L	D8EA or	DR-8ES
	X24ESR-U	X24ES-U	X27ESR-U
Gap	0.6 – 0.7 mm	0.6 – 0.7 mm	0.6 – 0.7 mm
	(0.024 – 0.028 in)	(0.024 – 0.028 in)	(0.024 to 0.028 in)

Torque settings

Component	kgf m	lbf ft
Automatic timing unit (ATU) bolt	2.1 – 2.5	15 – 18
Sparking plugs	1.2 – 1.9	9 – 14

1 General description

The Honda dohc models covered by this manual make use of capacitor discharge ignition (CDI) systems, in which the ignition timing and triggering is controlled electronically. As a result, the spark at the plugs is more powerful and is accurately timed. Because there are no mechanical aspects to the ignition system, wear does not take place, and thus the ignition timing will remain accurate unless disturbed, or in the rare event of component failure.

Like most four cylinder machines, the Honda dohc models employ what is effectively two ignition systems, each controlling two cylinders. This means that each time a spark is generated in one of the ignition coils, it is fed to two of the sparking plugs. Combustion will only take place in the cylinder which is under compression, leaving one spark wasted. This arrangement is known as the 'spare spark' system. It is important to remember this when dealing with ignition system faults, as these are most likely to be confined to one half of the system. The system can therefore be divided into that which relates to cylinders 1 and 4, and the corresponding equipment relating to cylinders 2 and 3.

The system is triggered by a reluctor mounted on the crankshaft end which rotates past two pickup coils or pulsers. The reluctor is analogous to the contact breaker cam in a conventional system. As the peak of the reluctor passes the pulser coil, it generates a small signal current. This current triggers the appropriate circuit in the ignition amplifier, or spark unit. The high voltage pulse from the amplifier feeds to the relevant coil, where its passage through the coil's primary windings induces the required high tension (HT) spark in the secondary windings. The reluctor continues to rotate, causing the same chain of events to take place in the remaining circuit.

For this range of models, Honda have chosen to retain a mechanical automatic timing unit (ATU) rather than opt for full electronic advance. As the engine speed rises, small weights on the ATU are thrown outwards against spring pressure. This movement is translated to the reluctor, which advances in relation to the crankshaft, thus providing an advanced spark for high speed running.

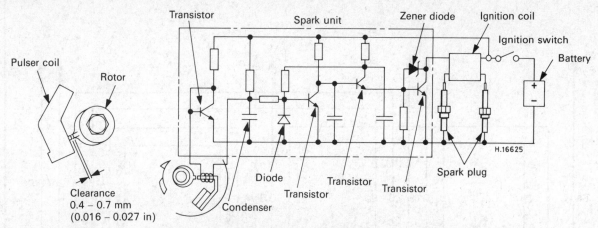

Fig. 3.1 Ignition sub-system

Note that the above system is effectively duplicated, each sub-system controlling a pair of spark plugs.

2 Electronic ignition system: testing

1 The CDI system, as mentioned previously, can be divided into two separate systems for most fault diagnosis purposes. Total failure of the entire system is unlikely, and trouble is usually confined to 1 and 4 cylinder or to 2 and 3 cylinders. In this case, it can be assumed that the system is functional up to the amplifier, or spark unit.

2 A preliminary check may be made as follows. Remove the sparking plugs from cylinders 1 and 2, remembering that these each represent one sub-system. Remove the inspection cover from the left-hand side of the unit to reveal the pulser coils and reluctor. Check that the engine kill switch and the main ignition switch are on.

3 Using a screwdriver with an insulated handle, bridge the gap between the reluctor and the metal core of one of the pulser coils. If the ignition circuitry is sound a spark will occur at the appropriate plug. Repeat the check with the remaining pulser and plug. If there is a defect in the system one of the plugs will fail to spark.

4 If no spark at all was found, attention should be turned to the ignition switch. Check that the switch operates properly, and that power is being fed to the ignition amplifier (spark unit). Ensure that the kill switch is at the 'run' position and has not shorted. If all seems well, and the remaining wiring connections are sound, it can be assumed that the amplifier is in need of attention. As it is unlikely that both pulsers, both coils or all four sparking plugs will have failed simultaneously, they can be ignored for the time being.

5 If one of the plugs failed to produce a spark in the above test, the system can be assumed to be sound up to the amplifier unit. The latter should be checked, followed by the pulser coil(s) and ignition coil(s).

6 With the ignition switched off, check the clearance between the steel core of the pulser coils and the raised section of the reluctor using feeler gauges. If the clearance is outside the range 0.4 – 0.7 mm (0.016 – 0.027 in) adjust the position of the pulser coil on the baseplate to bring it within this range.

3 Ignition amplifier (spark unit): testing

1 If preliminary checks have indicated a possible fault in the amplifier unit it should be tested as described below. Note that a dc voltmeter, or a multimeter set on the appropriate scale, will be required for the test. It is assumed that owners with access to this equipment will be conversant with its use. If this is not the case, the work should be entrusted to a Honda Service Agent or an Auto-electrician.

Test A

2 Trace the pulser leads back to the connector block behind the left-hand side cover. These terminate in a red 6-pin connector block which is mounted at the bottom of the connector bracket. Disconnect the pulser leads. Connect a multimeter, set on 0-50 volts dc, as follows. Attach the positive (+) probe to the blue wire at the white 4-pin connector. Connect the negative (–) probe of the multimeter to earth. Using a length of wire as a jump lead, connect one end to the blue lead with the white sleeve at the wiring harness side of the red 6-pin connector. Switch on the ignition, and flash the end of the jump lead to earth. This operation should result in the meter needle fluctuating between 12V and 0V if the unit is working correctly.

Test B

3 Repeat the above test sequence but this time with the positive multimeter probe connected to the yellow lead of the other white 6-pin connector, and the jump lead attached to the yellow/white sleeve lead at the red connector, thus testing the 2-3 cylinder amplifier unit. Similar results should be obtained in both tests. If the unit(s) prove faulty, they must be renewed since there is no practicable form of repair. Refer to the accompanying circuit diagram for details of the test connections.

4 Pulser coils: testing

1 The pulser coils can fail in two possible ways; either by an internal break in the windings (open circuit) or by an internal failure of the winding insulation (short circuit). A pulser in good condition will show a specific resistance reading when checked as described below.

2 Disconnect the pulser leads at the red 6-pin connector behind the left-hand side cover. Using an ohmmeter or a multimeter set on the resistance scale, measure the resistance between the yellow leads (2-3 cylinders) and then the blue leads (1-4 cylinders). In each case a resistance of 530 ± 50 ohms at 20°C (68°F) should be indicated. Open or short circuits will be indicated by infinite or zero resistance respectively. A defective pulser must be renewed – there is no satisfactory means of repair.

3 Measure the pulser coil/reluctor air gap, ie the clearance between the tooth of the reluctor and the tooth of the pulser coil (see Fig. 3.1 inset). The correct clearance should be 0.4 – 0.7 mm (0.016 – 0.027 in); slacken the retaining screws and move the coil accordingly if adjustment is required. Repeat the procedure on the other coil.

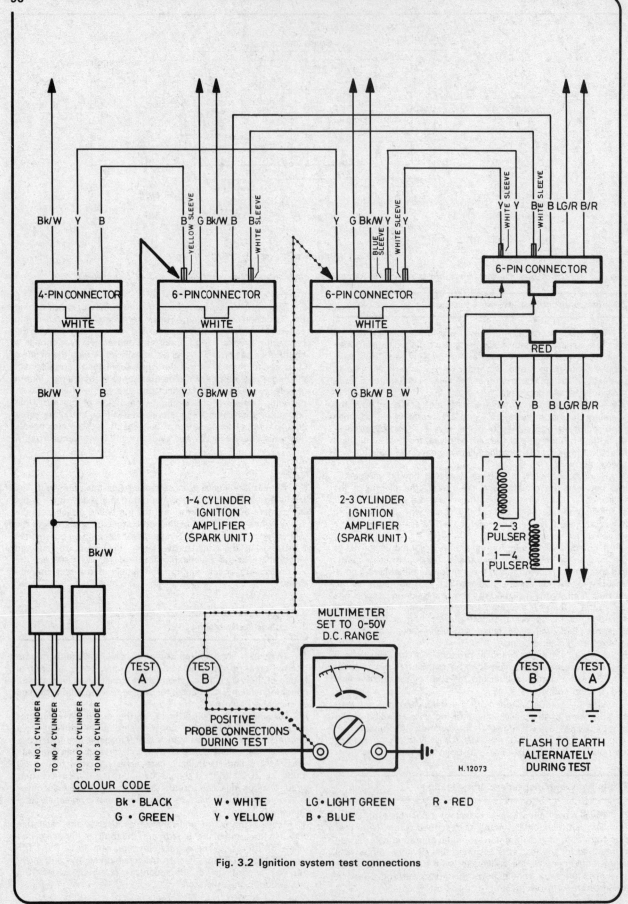

Fig. 3.2 Ignition system test connections

3.1 A: 1-4 spark unit connector B: 2-3 spark unit connector
C: Coil connector   D: Pulser connector (red)

3.2 Separate red connector during tests

4.1 Ignition pulser assembly

5 Automatic timing unit (ATU): examination

1   The automatic timing unit rarely gives rise to problems unless of considerable age or if neglect has allowed the pivots to become corroded. As CDI ignition is fitted attention to this area is likely to be far from frequent. Although not specified as a routine servicing task, it is worth checking the unit annually to reduce the chance of failure. Access to the unit is gained by removing the complete front left-hand engine casing, together with the ignition pickup assembly. The operation does not require that the ignition timing be lost.
2   The unit comprises spring loaded balance weights, which move outward against the spring tension as centrifugal force increases. The balance weights must move freely on their pivots and be rust-free. The tension springs must also be in good condition. Keep the pivots lubricated and make sure the balance weights move easily, without binding. Most problems arise as a result of condensation, within the engine, which causes the unit to rust and balance weight movement to be restricted.
3   The automatic timing unit mechanism is fixed in relation to the crankshaft by means of a dowel. In consequence the mechanism cannot be replaced in anything other than the correct position. This ensures accuracy of ignition timing to within close limits.
4   The correct functioning of the auto-advance unit can be checked when the engine is running by the use of a

stroboscopic light. If a strobe light is available, connect it to the ignition circuit of the 1 and 4 cylinders as directed by the manufacturer of the light. With the engine running, direct the beam of light at the fixed timing mark on the crankcase, through the aperture in the base plate. At tickover the timing mark and the **1,4 F-I** mark on the auto-advance unit should be precisely aligned. When the engine is running at 6000 rpm or above, the timing mark should align with two parallel lines which are marked on the automatic timing unit slightly in advance of the 'F' mark. The above test relies, of course, on the static ignition timing being correct.
5   If the unit proves to be faulty it must be renewed if cleaning and lubrication fails to effect a repair. No replacement parts are available to overhaul the unit.

6 Ignition coils: location and testing

1   Two separate ignition coils are fitted, each of which supplies a different pair of cylinders. The coils are mounted below the petrol tank, each side of the frame top tube.
2   If a weak spark, poor starting or misfiring causes the performance of the coils to be suspected, they should be tested by a Honda Service Agent or an auto-electrician who will have the appropriate test equipment.
3   It is unlikely that the coils will fail simultaneously. If intermittent firing occurs on one pair of cylinders the coils may be swopped over by interchanging the low tension terminal leads and the HT leads. If the fault then moves from one pair of cylinders to the other, it can be taken that the coil is faulty.
4   The coils are sealed units and therefore if a failure occurs, repair is impracticable. The faulty item should be replaced by a new component.

7 Ignition timing: checking

1   The ignition timing is set accurately at the manufacturing stage and in the normal course of events will maintain its accuracy for an indefinite period. In the event that the timing setting has been lost in the course of overhaul, it can be checked either statically or dynamically, as described below.
2   The timing is best checked dynamically, using a stroboscope timing lamp. Start by removing the pulser inspection cover. Following the maker's instructions, connect the timing lamp to No 1 cylinder's high tension lead. Start the engine and allow it to idle. If the lamp is directed at the timing aperture in the pulser base plate, it will be noted that it appears to 'freeze' the timing marks on the rotating ATU. At the normal idle speed of 1000 ± 100 rpm, the **1.4 F-I** mark (the scribed line

to the left of the letter I) should be aligned with the fixed index mark.

3   If adjustment is required, slacken the pulser base plate screws. These are situated at the 6 o'clock and 12 o'clock positions. The base plate can now be rotated to set the timing position. Tighten the screws and re-check the timing as described above.

4   If a stroboscopic timing lamp is not available, the timing may be checked conveniently, if less accurately, as a static setting. Using the large hexagon on the crankshaft end, rotate the latter until the 1.4 S-F mark lines up with the index mark, ie the scribed line to the right of the letter S. If the reluctor is viewed end-on, the projecting lug should align with a similar projection at the centre of the appropriate pickup or pulser coil. If necessary, adjust the pulser baseplate until the two coincide.

8   Sparking plugs: checking, cleaning and resetting the gaps

1   The Honda 750 and 900 dohc models are fitted with NGK or ND sparking plugs of various types. These are described in the specifications Section at the beginning of the Chapter. Note that in certain operating conditions, a change of plug grade may be required, through the standard grade will prove adequate in most cases.

2   Check the plugs points gap every 4000 miles, renewing the plugs every 8000 miles. This is the recommended maintenance interval given by the manufacturer, but reducing it somewhat would not be harmful. To reset the gap, bend the outer electrode closer to or further away from the central electrode until the gap is within the range 0.6 – 0.7 mm (0.024 – 0.028 in) as measured with a feeler gauge. The gap is usually set to the lower figure to allow for erosion of the electrodes. Never bend the centre electrode or the insulator will crack, causing engine damage if the particles fall into the cylinder whilst the engine is running.

3   With some experience, the condition of the sparking plug electrodes and insulator can be used as a reliable guide to engine operating conditions.

4   Always carry a spare sparking plug of the recommended grade. In the rare event of plug failure, this will allow immediate replacement and prevent nuisance and the strain on the engine when attempting to continue on three cylinders.

5   If the threads in the cylinder head strip as a result of over tightening the sparking plugs, it is possible to reclaim the head by means of a Helicoil thread insert. This is a cheap and convenient method of replacing the threads; most motorcycle dealers operate a service of this nature at an economical price.

6   Make sure the plug insulating caps are a good fit and have their rubber seals. They should be kept clean to prevent leakage and tracking of the HT current. These caps contain the suppressors that eliminate both radio and TV interference.

5.1 ATU is mounted on crankshaft LH end

5.2 Check unit for wear. Note timing marks

6.1 Coils are mounted beneath top frame tubes

Electrode gap check - use a wire type gauge for best results

Electrode gap adjustment - bend the side electrode using the correct tool

Normal condition - A brown, tan or grey firing end indicates that the engine is in good condition and that the plug type is correct

Ash deposits - Light brown deposits encrusted on the electrodes and insulator, leading to misfire and hesitation. Caused by excessive amounts of oil in the combustion chamber or poor quality fuel/oil

Carbon fouling - Dry, black sooty deposits leading to misfire and weak spark. Caused by an over-rich fuel/air mixture, faulty choke operation or blocked air filter

Oil fouling - Wet oily deposits leading to misfire and weak spark. Caused by oil leakage past piston rings or valve guides (4-stroke engine), or excess lubricant (2-stroke engine)

Overheating - A blistered white insulator and glazed electrodes. Caused by ignition system fault, incorrect fuel, or cooling system fault

Worn plug - Worn electrodes will cause poor starting in damp or cold weather and will also waste fuel

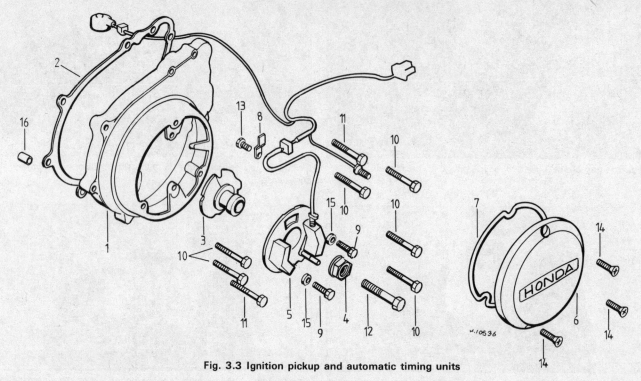

Fig. 3.3 Ignition pickup and automatic timing units

1   Crankcase cover
2   Gasket
3   Automatic timing unit
4   Engine turning hexagon
5   Pulser assembly
6   Cover
7   Seal
8   Timing index pointer
9   Bolt – 2 off
10 Bolt – 6 off
11 Bolt – 2 off
12 Bolt
13 Bolt
14 Oval headed screw – 3 off
15 Washer – 2 off
16 Dowel pin – 2 off

9   Fault diagnosis: ignition system

Symptom	Cause	Remedy
Engine will not start	Faulty ignition switch	Operate switch several times in case contacts are dirty. If lights and other electrics function, switch may need renewal
	Faulty CDI unit	Have unit tested. Replace if required
	Wiring fault	Check and repair wiring
	Starter motor not working	Discharged battery. Remove battery from machine and recharge
		Faulty starter circuit. Check for continuity
	Short circuit in wiring	Check whether fuse is intact. Eliminate fault before switching on again
	Completely discharged battery	If lights do not work, remove battery and recharge
Engine misfires	Fouled sparking plug	Renew plug and have original cleaned
	Pulser failure	Check pulsers, renew if required
Engine lacks power and overheats	Retarded ignition timing due to ATU unit failure	Check ATU
Engine 'fades' when under load	Pre-ignition	Check grade of plugs fitted; use recommended grades only

Chapter 4 Frame and forks

For modifications, and information relating to later models, see Chapter 7

Contents

Specifications

Front forks

	750 models	900 model
Type	Oil damped telescopic	
Fork spring free length	507.8 mm (20.0 in)	551.0 mm (21.7 in)
Service limit	492.6 mm (19.4 in)	540.0 mm (21.3 in)
Fork stanchion OD	34.93 – 34.95 mm	34.90 – 34.96 mm
	(1.3752 – 1.3760 in)	(1.3740 – 1.3764 in)
Service limit	34.90 mm (1.374 in)	34.85 mm (1.372 in)
Lower leg ID	35.042 – 35.104 mm	35.042 – 35.104 mm
	(1.3976 – 1.3820 in)	(1.3976 – 1.3820 in)
Service limit	35.15 mm (1.384 in)	35.25 mm (1.388 in)
Max stanchion bend	0.2 mm (0.01 in)	0.2 mm (0.01 in)
Fork oil capacity (per leg):		
Dry	172.5 – 177.5 cc	175 cc
At oil change	155cc	155 cc
Fork oil grade	Automatic transmission fluid (ATF)	

Rear suspension

	CB750 K(Z)	CB750 LTD	CB750 F	CB900 F
Type	Swinging arm, gas/oil rear suspension units			
Spring preload adjustment	5 position			
Compression damping adjustment	Fixed	Fixed	pre 1980; Fixed 1980 on; 2 position	2 position
Rebound damping adjustment	Fixed	Fixed	pre 1980; Fixed 1980 on; 3 position	3 position
Spring free length	244.5 mm (9.6 in)	244.5 mm (9.6 in)	pre 1980; 244.5 mm (9.6 in) 1980 on; 238 mm (9.37 in)	238.0 mm (9.37 in)
Service limit	237.2 mm (9.5 in)	237.2 mm (9.5 in)	pre 1980; 237.2 mm (9.5 in) 1980 on; 233 mm (9.17 in)	233.0 mm (9.17 in)

Swinging arm bearing:		
Type	pre 1980; Bush / 1980 on; Needle roller	pre 1980; Bush / 1980 on; Needle roller
Swinging arm bush ID	21.500 – 21.552 mm (0.8465 – 0.8485 in)	21.500 – 21.552 mm (0.8465 – 0.8485 in)
Service limit	21.7 mm (0.854 in)	21.7 mm (0.854 in)
Inner sleeve OD	21.427 – 21.460 mm (0.8436 – 0.8449 in)	21.427 – 21.460 mm (0.8436 – 0.8449 in)
Service limit	21.4 mm (0.843 in)	21.4 mm (0.843 in)

Torque settings

Component	kgf m	lbf ft
Steering stem nut	8.0 – 12.0	58 – 87
Handlebar clamp:		
750 models	1.8 – 2.5	13 – 18
900 model	2.8 – 3.2	20 – 23
Fork clamp bolts (upper)	0.9 – 1.3	7 – 9
Fork clamp bolts (lower):		
750 models	3.0 – 4.0	22 – 29
900 model	1.8 – 2.5	13 – 18
Fork cap bolt	2.0 – 3.0	15 – 22
Wheel spindle clamp nuts	1.8 – 2.5	13 – 18
Wheel spindle nut	5.5 – 6.5	40 – 47
Swinging arm pivot	5.5 – 7.0	40 – 51
Steering head adjuster nut:		
750 models	0.8 – 1.2	6 – 9
900 model	0.2 – 0.4	1.5 – 2.9
Gearchange pedal	0.8 – 1.2	6 – 9
Rear suspension unit (upper):		
750 models	2.0 – 3.0	14 – 22
900 model	3.0 – 4.0	22 – 29
Rear suspension unit (lower)	3.0 – 4.0	22 – 29

1 General description

The Honda 750 and 900 dohc models employ a welded tubular steel frame of conventional full cradle design. The right-hand lower cradle section can be removed to permit engine removal.

The front forks are hydraulically damped, consisting of two telescopic shock absorber assemblies, each of which comprises an inner tube, an outer tube, a spring and a cylinder, piston and valve. The whole fork assembly is attached to the frame by the steering head stem and is mounted on two bearing assemblies contained in the steering hea housing.

The damping action of the fork is accomplished by the flow resistance of the fork oil flowing between the inner and outer tubes.

Rear suspension is by means of a pivoted rear fork or swinging arm, supported by a pair of gas filled, oil damped rear suspension units.

2 Front forks: methods of removal

1 If necessary, it is possible to remove the forks as an assembly, together with the lower yoke. It is unlikely that this method will prove advantageous in view of the amount of preliminary dismantling necessary, but if this approach is deemed essential, follow the sequence detailed in Section 4, leaving the stanchions clamped in the lower yoke.

2 In most cases, the forks are best removed individually, without disturbing the yokes or steering head bearings. The relevant sequence is described in Section 3 of this Chapter.

3 Front fork legs: removal and replacement

1 Place the machine securely on its centre stand, leaving adequate room around the front wheel area for comfortable working. It should be noted at this stage that some means of supporting the front wheel clear of the ground must be arranged. This can be accomplished by placing blocks or a jack beneath the sump, taking care not to damage the delicate sump fins. Alternatively a pair of tie-down straps, of the type used to secure motorcycles on trailers, can be arranged to lash the rear of the machine to the ground. Whatever method is chosen, make sure that there is no risk of the machine falling over whilst the front wheel and forks are removed.

2 On models equipped with twin front disc brakes, it will be necessary to release one of the calipers to permit front wheel removal. Note that if both legs are to be removed, both calipers should be released. This applies to single disc models where the left-hand fork leg is to be removed, even though wheel removal is possible with the caliper in place.

3 Slacken the caliper retaining bolts and lift the caliper clear of the forks. Insert a thin wooden wedge between the disc pads to prevent their displacement should the front brake lever be inadvertently operated prior to the calipers being refitted. The caliper(s) should be tied to the frame to avoid undue strain being placed on the hydraulic hose(s).

4 Release the speedometer cable by unscrewing the cross-head screw which secures it to the drive gearbox. Slacken the wheel spindle clamp nuts to release the wheel, which can now be lowered and disengaged from the front forks. Remove the four bolts which secure the front mudguard stays to the fork lower legs. Lift the mudguard clear and disengage the speedometer cable from its guide loop.

5   The CB900F model employs individual cast handlebar sections, these being clamped to the protruding ends of the fork stanchions. It follows that they must be released before the fork legs can be removed. Each is retained by a single clamp bolt and also a cap bolt which screw into the top of the stanchion. Note that the handlebar sections should be lodged clear of the fork area. One method is to place them on the fuel tank, having first protected the tank with rags to avoid paintwork damage.

6   The forks legs can now be removed after releasing the upper and lower pinch bolts. Note that if it is intended to dismantle the forks, the fork top bolt should be slackened prior to removal. This is not easy with the forks in their normal position, as the handlebar obstructs the bolts. The easiest method is to slacken the pinch bolts, slide the stanchion down by about 3 inches then tighten the lower yoke pinch bolt. Prise off the plastic cap, and slacken the top bolt using the appropriate hexagon socket key. The pinch bolt can now be released and the fork leg lowered clear of the yoke. Repeat the operation with the remaining fork leg, where required.

The fork legs can be refitted by reversing the above sequence, noting the following points. The fork top bolts should be tightened to 2.0 – 3.0 kgf m (15 – 22 lbf ft). This presents something of a problem unless a hexagon key socket is available, but can be overcome by sawing a short section off an ordinary key, and using a torque wrench and socket to drive it. Do not omit to fill the forks with the recommended quantity (172.5 – 177.5 cc) of automatic transmission fluid (ATF) per leg.

7   Fit each leg so that the edge of the plain face of each stanchion is flush with the top yoke. Tighten the lower yoke pinch bolts to 3.0 – 4.0 kgf m (22 – 29 lbf ft) followed by the upper yoke pinch bolts to 0.9 – 1.3 kgf m (7 – 9 lbf ft). When fitting the front wheel, note that the wheel spindle clamps are marked with an arrow and a letter F, denoting front. The front clamp nut should be tightened to the torque figure given below, followed by the rear nut, leaving a small gap to the rear of the spindle. The correct torque figure is 1.8 – 2.5 kgf m (13 – 18 lbf ft).

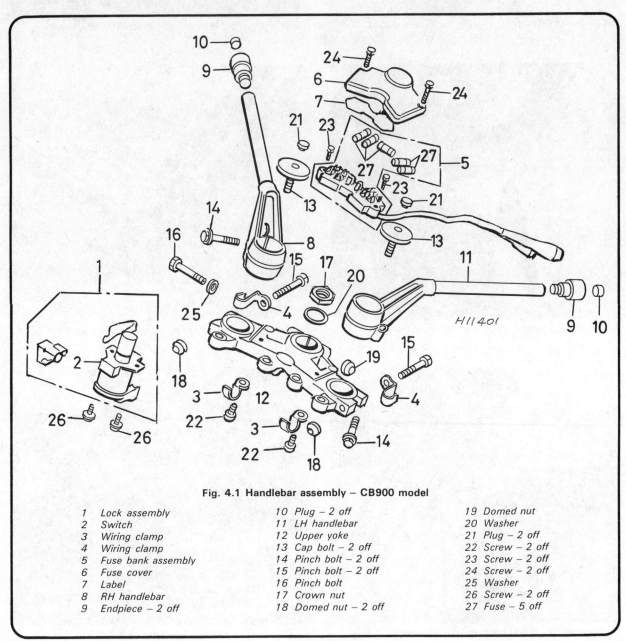

Fig. 4.1 Handlebar assembly – CB900 model

1 Lock assembly	10 Plug – 2 off	19 Domed nut
2 Switch	11 LH handlebar	20 Washer
3 Wiring clamp	12 Upper yoke	21 Plug – 2 off
4 Wiring clamp	13 Cap bolt – 2 off	22 Screw – 2 off
5 Fuse bank assembly	14 Pinch bolt – 2 off	23 Screw – 2 off
6 Fuse cover	15 Pinch bolt – 2 off	24 Screw – 2 off
7 Label	16 Pinch bolt	25 Washer
8 RH handlebar	17 Crown nut	26 Screw – 2 off
9 Endpiece – 2 off	18 Domed nut – 2 off	27 Fuse – 5 off

3.4 Remove mudguard by releasing four retaining bolts

3.6a Slacken clamp bolts and draw fork leg downwards

3.6b It is worth removing plug and cap bolt at this stage

3.6c Bolt can be used as improvised hexagon wrench

3.6d Reassemble in the reverse order of dismantling

3.6e Top up oil before fork is in final position

4 Fork yokes and steering head bearings: removal and replacement

1   Fork yoke dismantling is an unwieldy process which is best avoided if at all possible. A considerable amount of work is involved in removing the numerous appendages from the fork yokes. For this reason, fork removal is best carried out as described in Section 3. It should be noted that some operations, such as inspection of the steering head bearings, can be undertaken without resorting to the full strip-down sequence given here, the trick being to disturb only those components which directly affect the part-dismantling sequence required. Before disconnecting any electrical connections the battery must be isolated by removing the lead from the positive terminal.

2   Start by removing the front wheel and fork legs as described in Section 3 of this Chapter. Although it is possible to leave the fork legs attached to the lower yoke, this makes an awkward job all the more difficult; they are best removed. The handlebar assembly must be removed, either completely after releasing the various switch harnesses, hydraulic pipe and cables, or by removing the handlebar clamp and displacing the assembly to provide clearance. In each case it is best to remove the fuel tank to avoid damage to the paintwork.

3   On all but the CB900F model, the handlebar is secured by a wide clamp which doubles as a fuse holder. Remove the fuse cover to expose the four socket screws. These should be released and the fuse holder displaced to release the handlebar. Move the handlebar assembly rearward to clear the yoke, lodging it so that the master cylinder reservoir is not likely to leak. In the case of the CB900F, the individual handlebars are clamped to the top of the fork stanchions, and will have been removed during the fork removal sequence.

4   If it is wished to remove the entire handlebar assembly, it will be necessary to release the handlebar switches, control cables and levers and the brake master cylinder. This is unlikely to be necessary in normal circumstances, except where accident damage has necessitated a full stripdown and examination of the frame, steering and suspension components.

5   Slacken and remove the screws which secure the headlamp assembly to the headlamp shell. Lift the unit away, disconnecting the bulb connector(s). The shell houses a number of multi-pin electrical connectors. These enter through a number of holes in the rear of the shell and will require separation before the shell can be released. Unscrew the reflectors to expose the headlamp shell retaining bolts, which can then be removed to permit headlamp shell removal.

6   Remove the instrument assembly by releasing the rubber-mounted retaining nuts and detaching the speedometer and tachometer drive cables. Lift the assembly clear together with the wiring harness and connector block. The head-lamp/instrument mounting bracket can now be removed.

7   Work can now start on dismantling the steering head and yoke assembly. Note that some careful manoeuvring will be required to free the upper and lower yokes if the handlebar assembly has been displaced but not fully dismantled. Remove the steering stem cap nut, noting that Honda supply a special 30-32 mm double-ended socket, part number 07716-0020400, if required.

8   The upper yoke can now be removed, tapping it upwards to free it from the steering stem. With the upper yoke removed, the slotted adjustment nut will be exposd, and can be slackened using a C-spanner. As the nut is removed, the lower yoke and steering stem will drop downwards, and thus should be supported until the nut is clear. Carefully disengage the lower yoke assembly and place it to one side. Reassembly is a reversal of the above sequence, noting that the steering head bearings should be checked, greased and adjusted as described in Section 5.

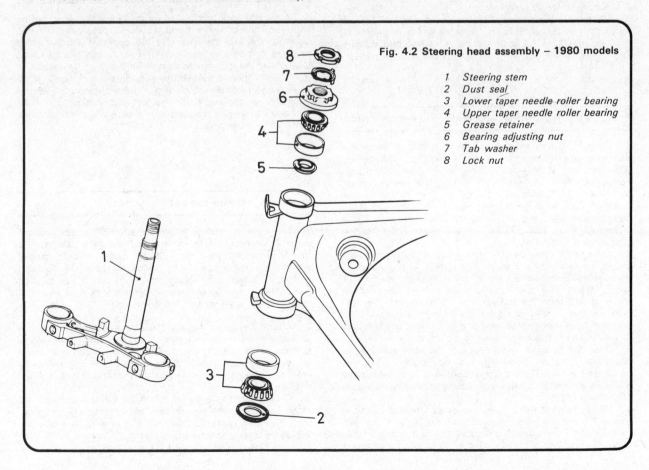

Fig. 4.2 Steering head assembly – 1980 models

1   Steering stem
2   Dust seal
3   Lower taper needle roller bearing
4   Upper taper needle roller bearing
5   Grease retainer
6   Bearing adjusting nut
7   Tab washer
8   Lock nut

5   Steering head bearings: examination and adjustment

1   The steering head bearings are of the taper roller variety, and are unlikely to give rise to problems in the normal life of the machine, although the manufacturer recommends examination, lubrication and adjustment of these components at 8000 mile intervals. Access to the bearings is gained by following the procedure detailed in Section 4 of this Chapter.

2   It will be noted that the bearings are effectively in two parts; the outer races, which will remain in the steering head tube, and the inner race, cage and rollers, which will come away as the lower yoke is removed. Of the latter, the lower bearing will probably be firmly attached to the steering stem, whilst the upper bearing will lift away quite easily. It is normally possible to lever the lower bearing off the steering stem, but it may prove necessary to employ a bearing extractor in stubborn cases.

3   The outer races can be removed with the aid of a long drift, passed through the steering head tube, new races being fitted by judicious use of a tubular drift, such as a large socket or similar. Before any decision is made to remove the outer races, they should be cleaned and checked as described below.

4   Wash out the bearings with clean petrol to remove all traces of old grease or dirt. Check the faces of the rollers and the outer races for signs of wear or pitting, both of which are unlikely unless the machine has been neglected in the past. If damaged, the bearings must be renewed.

5   When assembling the steering head, pack each bearing with grease prior to installation. The lower yoke is offered up and the upper bearing and adjuster nut is fitted finger tight. Adjustment of the bearings requires a socket-type peg spanner so that the slotted adjuster nut can be set to the prescribed figure. This requires the use of the special Honda steering stem socket (Part No 07916-3710100) or a home-made equivalent. A piece of tubing can be filed to fit the nut and then welded to a damaged socket to improvise.

6   Tighten the adjuster nut, first to 3.0 – 4.0 kgf m (22 – 29 lbf ft), then slacken it off again. Reset the torque wrench to 10 – 20 kgf cm (0.7 – 1.4 lbf ft). this being the final torque figure. This sequence ensures that the bearings bed down squarely.

6   Front fork legs: dismantling, renovation and reassembly

1   Having removed the fork legs as described in Section 3, they may be dismantled for further examination. Always deal with one leg at a time and on no account interchange components from one leg to the other as the various moving parts will have bedded in during use, and should remain matched. Commence by draining the oil, either by way of the drain plug in the lower leg or by removing the top bolt and inverting the leg. Pumping the unit will assist in the draining operation.

2   Refer to the accompanying line drawing then start dismantling, laying each part out on a clean surface as it is removed. Remove the chromium-plated top bolt and withdraw the fork spring. Using a 6 mm hexagon key, release the damper retaining bolt from the underside of the lower fork leg. It is helpful to clamp the lower leg in a vice during this operation, but care should be taken not to damage the soft alloy. Use soft jaws or wrap some rag around the leg to protect it.

3   Once the bolt has been released, the stanchion can be withdrawn from the lower leg. The alloy damper seat may remain in the lower leg, and should be shaken free. The damper assembly will remain inside the stanchion and can be tipped out of the upper end.

4   The parts most liable to wear over an extended period of service are the internal surfaces of the lower leg and the outer surfaces of the fork stanchion or tube. If there is excessive play between these two parts they must be replaced as a complete

unit. Check the fork tube for scoring over the length which enters the oil seal. Bad scoring here will damage the oil seal and lead to fluid leakage.

5   It is advisable to renew the oil seals when the forks are dismantled even if they appear to be in good condition. This will save a strip-down of the forks at a later date if oil leakage occurs. The oil seal in the top of each lower fork leg is retained by an internal C ring which can be prised out of position with a small screwdriver. Check that the dust excluder rubbers are not split or worn where they bear on the fork tube. A worn excluder will allow the ingress of dust and water which will damage the oil seal and eventually cause wear of the fork tube.

6   It is not generally possible to straighten forks which have been badly damaged in an accident, particularly when the correct jigs are not available. It is always best to err on the side of safety and fit new ones, especially since there is no easy means to detect whether the forks have been overstressed or metal fatigued. Fork stanchions (tubes) can be checked, after removal from the lower legs, by rolling them on a dead flat surface. Any misalignment will be immediately obvious.

7   The fork springs will take a permanent set after considerable usage and will need renewal if the fork action becomes spongy. The service limit for the total free length of each spring is given in the Specifications. Always renew them as a matched pair.

8   Fork damping is governed by the viscosity of the oil in the fork legs, normally ATF, and by the action of the damper assembly. Each fork leg holds 172.5 – 177.5 cc of damping fluid.

7   Steering head lock: maintenance

1   A security lock is mounted on the headstock, enabling the owner to immobilise the machine by locking the steering in one position. The lock consists of a key operated plunger which engages in a slot in the steering head. A small return spring disengages the lock mechanism when the key is released.

2   Maintenance is confined to keeping the lock lightly lubricated using light machine oil or one of the multipurpose aerosol lubricants. In the event that the lock malfunctions, it will be necessary to remove the body after unscrewing the cover plate and to fit a replacement unit.

8   Frame: examination and renovation

1   The frame is unlikely to require attention unless accident damage has occurred. In some cases, renewal of the frame is the only satisfactory remedy if the frame is badly out of alignment. Only a few frame specialists have the jigs and mandrels necessary for resetting the frame to the required standard of accuracy, and even then there is no easy means of assessing to what extent the frame may have been overstressed.

2   After the machine has covered a considerable mileage, it is advisable to examine the frame closely for signs of cracking or splitting at the welded joints. Rust corrosion can also cause weakness at these joints. Minor damage can be repaired by welding or brazing, depending on the extent and nature of the damage.

3   Remember that a frame which is out of alignment will cause handling problems and may even promote 'speed wobbles'. If misalignment is suspected, as a result of an accident, it will be necessary to strip the machine completely so that the frame can be checked, and if necessary, renewed.

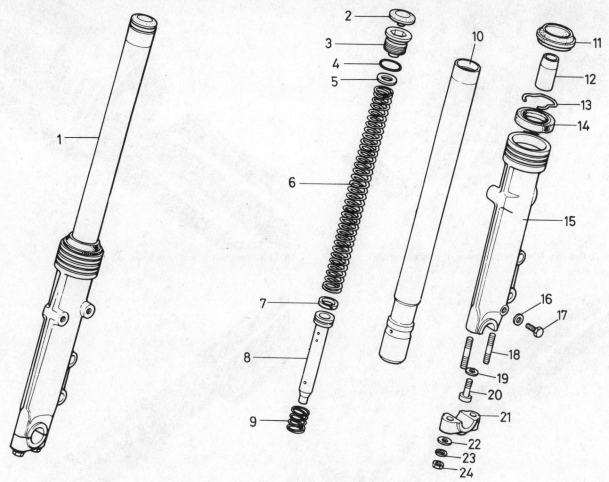

Fig. 4.3 Front fork – CB750K (other models similar)

1	Right-hand fork leg	9	Rebound spring – 2 off	17	Drain bolt – 2 off
2	Cap – 2 off	10	Stanchion – 2 off	18	Stud – 4 off
3	Top bolt – 2 off	11	Dust seal – 2 off	19	Washer – 2 off
4	O-ring – 2 off	12	Damper rod seat – 2 off	20	Bolt – 2 off
5	Washer – 2 off	13	Oil seal retainer – 2 off	21	Spindle clamp – 2 off
6	Spring – 2 off	14	Oil seal – 2 off	22	Washer – 4 off
7	Damper rod ring – 2 off	15	Left-hand lower leg	23	Spring washer – 4 off
8	Damper rod – 2 off	16	Sealing washer – 2 off	24	Nut – 4 off

6.2a Remove the fork top bolt ...

6.2b .. and withdraw the spring

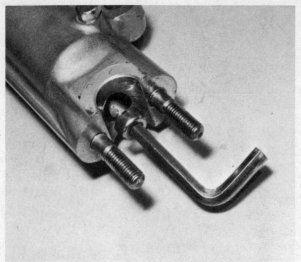

6.2c Remove Allen bolt to release damper assembly

6.3a Remove stanchion from lower leg

6.3b Damper assembly is arranged as shown

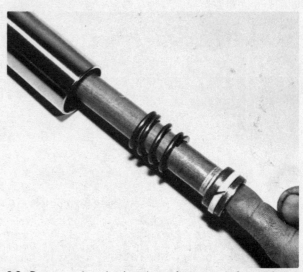

6.3c Damper rod can be tipped out of upper end of stanchion

6.4 Check condition of damper rod

6.5a Renew dust seal if damaged

6.5b Oil seal can be removed after dust seal and clip are freed

7.1 Steering lock is mounted on headstock

9 Swinging arm fork: removal and renovation

1 The swinging arm fork is supported on two headed bushed which pivot on an inner sleeve. The assembly is retained by a long pivot shaft which passes through lugs on the frame and through the centre of the sleeve. A grease nipple is fitted to enable grease to be pumped to the bearing surfaces.

2 Wear in the swinging arm bushes is characterised by a tendency for the rear of the machine to twitch when ridden hard through a series of bends. This can be checked by placing the machine on the centre stand, and pushing the swinging arm from side to side. Any discernible free play will necessitate the removal of the swinging arm for further examination.

3 Commence by detaching the silencer from each side of the machine, after slackening the clamp bolt and mounting nuts on each unit. In the case of models equipped with a four-into-four exhaust system, note that only the upper silencers can be detached. If further access is required the entire system must be removed from the machine. Detach the rear brake switch operating spring, then remove the brake adjusting nut. The brake operating rod can now be disengaged from the operating lever. Pull out the spring pin from the torque arm mounting stud, then remove the securing nut and disengage the torque arm. Release the chainguard mounting nuts, and lift the guard away.

4 Remove the split pin which retains the rear wheel spindle nut, then slacken the nut. Release and back off the chain tensioner drawbolts, and swing them through 90° degrees. The wheel can now be pushed forwards as far as possible, and the final drive chain disengaged from the rear wheel sprocket. Slide out the stops from the fork ends to allow the wheel assembly to be withdrawn rearwards. Manoeuvre the wheel clear of the frame and place it to one side to await reassembly.

5 Remove the lower suspension mounting bolt from each side of the machine, and push the units clear of the swinging arm. Slacken the pivot shaft nut and pull the pivot shaft out, supporting the swinging arm. The swinging arm can now be drawn rearwards, noting that the caps on each end of the pivot tube will probably drop clear. Disengage the fork from the drive chain, and place it on a bench to await further dismantling.

6 Displace the pivot sleeve and wash the sleeve and the headed bushes to remove all trace of grease. Examine the sleeve and bush bearing surfaces for signs of wear or scoring. If damaged or worn, or if below the limits given in the Specifications Section, replace the components as necessary. The two headed bushes may be driven out using a suitable bar or drift. When fitting new bushes, ensure that they are tapped squarely into the swinging arm bore, and that they seat securely. Clean

off any corrosion on the inner sleeve if this is to be re-used.

7 Reassemble in the reverse order of the above, lubricating the bushes with grease prior to installation. When assembly is complete, grease the pivot area by introducing grease via the grease nipple.

8 The 1980 models are equipped with needle roller bearings in place of the swinging arm bushes. This does not materially affect the dismantling sequence, though it should be noted that a Honda tool, No 07936-4250100 will be needed to effect removal of the bearings. Failing this, the bearings can be drifted out as described above. Note that in either case new bearings should be available, as the old items may well be destroyed during removal. It follows that the bearings should not be removed unless they are known to be worn out.

9 When fitting the new bearings, use Honda tool No 07946-4250100, or a suitable stepped drift. The latter must be a good fit in the bearing to prevent distortion. The bearings are fitted with the part numbered edge facing outwards. The bearings are followed by the shouldered thrust spacers. Pack the bearings with grease after installation.

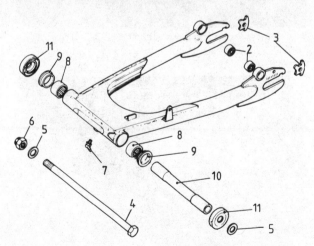

Fig. 4.4 Swinging arm – 1980 CB750K and CB750F

1 Swinging arm fork
2 Rubber bush – 2 off
3 Adjuster abutment – 2 off
4 Pivot bolt
5 Washer
6 Nut
7 Grease nipple
8 Bearing – 2 off
9 Shouldered collar
10 Centre spacer
11 Dust seal

9.3a Front chainguard mounting bolt

9.3b Rear chainguard mounting bolt

9.5a Withdraw swinging arm pivot bolt ...

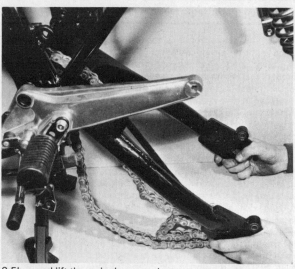

9.5b ... and lift the swinging arm clear

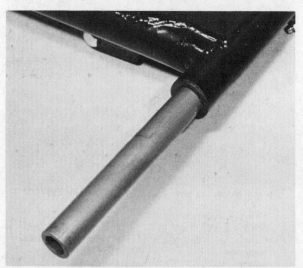

9.6a Remove sleeve and check for wear

9.6b Check condition of dust seals

9.7 A grease nipple is fitted at the centre of the swinging arm

10.1a Suspension units are retained by domed nut ...

10 Rear suspension units: examination and dismantling – all KZ and pre-1980 F models

1 The Honda 750 and 900 dohc models are fitted with hydraulically-damped rear suspension units featuring 5-way spring preload adjustment. Adjustment is effected by turning an annular cam near the bottom of the units. Little in the way of maintenance is possible, the gas/oil damper units being sealed. It is possible to remove the springs if a suitable spring compressor, Honda part No 07595-3290001 or equivalent, is available.

2 Assemble the spring compressor, checking that it is fitted securely with no risk of slippage. Compress the spring until the locknut on the underside of the top mounting lug can be slackened. Unscrew the top lug and decompress the spring. The spring can now be lifted off the damper body.

3 Check the damper unit for signs of leakage, noting that if traces of oil are discovered the damper must be renewed. It is not possible to dismantle or overhaul the units. Measure the free length of each of the suspension springs, renewing them as a pair if they have reached the service limit of 237.2 mm (9.5 in)

4 Reassemble in the reverse order of that described above, noting that the closest spring coils must face downwards. When fitting the locknut, apply a trace of thread locking compound. When refitting the units, tighten the upper mounting nuts to 2.0 – 3.0 kgf m (14 – 22 lbf ft). The lower bolts should be secured to 3.0 – 4.0 kgf m (22 – 29 lbf ft).

10.1b ... and by bolt at lower end

11 Rear suspension units: examination and dismantling – 1980 F models

1 The 1980 CB750F and CB900F models feature a more sophisticated version of the FVQ rear suspension unit. These units have the same spring preload adjustment of the earlier types, but also feature five damping adjustment settings. A slotted adjuster ring at the top of each unit provides three rebound damping settings, whilst a small lever at the base of each unit gives two compression damping settings. In each case, the adjuster serves to align valves with a range of oil bleed orifice sizes. This restricts the flow of damping oil to a greater or lesser extent, thus varying the damping rate.

2 The springs may be removed in a similar manner to that described in the preceding Section. Note that the upper shroud must be protected by wrapping it with pvc tape to prevent the polished surface becoming damaged by the spring compressor. Once compressed, the top adjusting ring can be removed, there being a slot in it to clear the damper rod. Note that the ring has

10.1c Check bonded rubber bushes, and renew as required

a driving slot which must engage with a tang on the damper rod during assembly.

3   Measure the free length of the spring. This should be at least 233.0 mm (9.17 in). If less than the above figure, renew the springs as a pair. No further dismantling is possible, the damper being a sealed assembly. Refit the spring with the closely-wound coils towards the bottom of the damper.

4   Honda recommend the following rebound and compression damping permutations as a guide. These may require some modification to suit particular road consitions or rider preference. Note that it is imperative that both units are set up similarly.

FVQ suspension adjustment settings

Conditions	Rebound setting	Compression setting
Rider only. Normal town roads	1	1
Rider only. Fast or twisting roads	2	1
Rider only. Poor road surface	3	1
Rider/passenger. Normal town roads	1	2
Rider/passenger or luggage. Fast or twisting roads	2	2
Rider/passenger or luggage. Poor road surface	3	2

12 Centre stand: examination

1   The centre stand pivots on a support tube which is clamped beneath the frame. A split pin is fitted at the clamped end of the tube as an additional security measure. A return spring retracts the stand when not in use. The assembly should be checked for wear or damage, and the pivot tube greased. It is important that the stand remains secure as a failure would allow quite a lot of damage to be incurred if the machine should fall over. More importantly, the effects of a stand becoming displaced whilst the machine is being ridden would be catastrophic. This applies to the return spring, which should be renewed promptly if showing any signs of abrasion or serious corrosion.

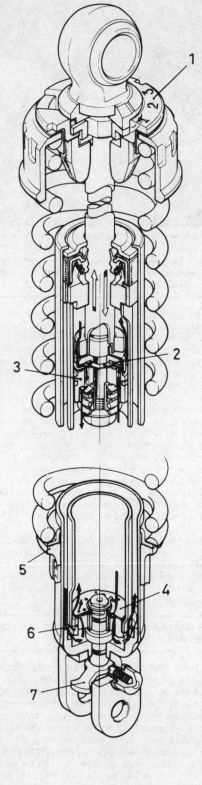

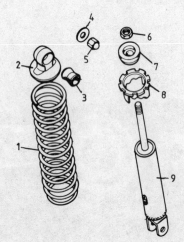

Fig. 4.5 Rear suspension unit – Early CB750K

1   Coil spring – 2 off
2   Top mounting – 2 off
3   Rubber cushion – 2 off
4   Washer – 2 off
5   Nut – 2 off
6   Nut – 2 off
7   Bump stop – 2 off
8   Spring adjuster – 2 off
9   Damper – 2 off

Fig. 4.6 Rear suspension unit – 1980 CB750K and CB750F

1   Upper adjusting ring
2   Rebound valve
3   Piston
4   Compression valve
5   Spring adjuster
6   Base valve
7   Lower adjusting lever

13 Prop stand: examination

1   The prop stand is secured to a plate on the frame with a bolt and nut, and is retracted by a tension spring. Make sure the bolt is tight and the spring is not overstretched, otherwise an accident can occur if the stand drops during cornering.

14 Footrests and rear brake pedal: examination

1   The footrests are of the swivel type and are retained by a clevis pin secured by a split pin. The advantage of this type of footrest is that if the machine should fall over the footrest will fold up instead of bending.
2   The rear brake pedal is held in position by a pinch bolt arrangement, the pedal return spring must be detached to remove the brake lever.

15 Dualseat: removal and replacement

1   The dualseat is attached to the frame by two clevis pins that are located with split pins on the right-hand side of the frame. To remove the seat, release the spring loaded catch on the right-hand side, and prop up the seat. Withdraw the two split pins from the clevis pivot pins, and remove the pivot pins. The seat mountings and damper rubbers can be left in place as the seat is lifted off.
2   If the dualseat is removed because it is torn, it is possible in most cases to find a specialist firm that recovers dualseats for an economical price, usually considerably cheaper than having to buy a new replacement. The usual charge is about 50% the cost of a new replacement, depending on the extent of the damage.

16 Instrument panel: removal and examination

1   The speedometer and tachometer heads form part of an instrument panel which also incorporates separate trip and total odometers (mileometers) and the various warning lamps. To remove the panel, it will first be necessary to remove the headlamp reflector unit so that the instrument wiring harness can be detached at the connector block. Release the speedometer and tachometer drive cables, then unscrew the two rubber-mounted retaining bolts from the underside of the panel. The panel can now be lifted clear and placed to one side.
2   Remove the four screws which retain the top of the panel, and lift this away. Turn the assembly over and release the two cap nuts to release the bottom section of the panel. The instruments are now retained by a single nut each to the main support bracket. Note that the instruments should not be left inverted as the damping fluid in each may start to leak out.
3   The speedometer and tachometer heads cannot be repaired by the private owner, and if a defect occurs a new instrument has to be fitted. Remember that a speedometer in correct working order is required by law on a machine in the UK also many other countries.
4   Speedometer and tachometer cables are only supplied as a complete assembly. Make sure the cables are routed correctly through the clamps provided on the top fork yoke, brake branch pipe, and frame.
5   Before an instrument head is condemned, check that the drive cable is not broken or kinked. In the latter case, the jerkiness produced will result in wildly inaccurate meter readings, even though the instrument itself may be quite functional.

17 Speedometer and tachometer drives: location and examination

1   The speedometer is driven from a gear inside the front wheel hub assembly. The gear is driven internally by a tongued washer (receiver). The receiver engages with two slots in the wheel hub, on the left-hand side. As the whole gearbox is prepacked with grease on assembly it should last the life of the machine, or until new parts are fitted. The spiral pinion that drives off the internal gear is retained in the speedometer gearbox casing by a grub screw, which should always be secured tightly.
2   The tachometer drive runs off the camshaft in the cambox and screws directly into the cylinder head cover in the centre position. The cable is retained by a screwed ferrule, in the same manner as the speedometer cable.

14.1 Footrests are retained by clevis pin and split pin

15.1 Seat hinge is retained as shown here

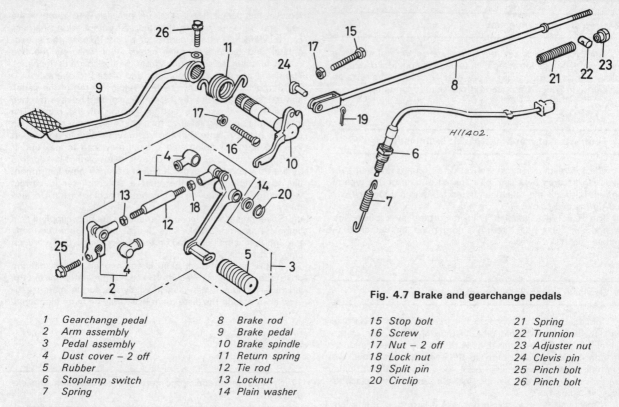

Fig. 4.7 Brake and gearchange pedals

1	Gearchange pedal	8	Brake rod
2	Arm assembly	9	Brake pedal
3	Pedal assembly	10	Brake spindle
4	Dust cover – 2 off	11	Return spring
5	Rubber	12	Tie rod
6	Stoplamp switch	13	Locknut
7	Spring	14	Plain washer

15	Stop bolt	21	Spring
16	Screw	22	Trunnion
17	Nut – 2 off	23	Adjuster nut
18	Lock nut	24	Clevis pin
19	Split pin	25	Pinch bolt
20	Circlip	26	Pinch bolt

18 Cleaning the machine

1   After removing all the surface dirt with warm water and a rag or sponge, use a cleaning compound such as 'Gunk' or 'Jizer' for the oily parts. Apply the cleaner with a brush when the parts are clogged so that it has an opportunity to soak into the film of oil or grease.

Finish off by washing down liberally, taking care that water does not enter into the carburettors, air cleaner or electrics. If desired, a polish such as Solvol Autosol can be applied to the alloy parts to give them a full lustre, Application of a wax polish to the cycle parts and a good chrome cleaner to the chrome parts will also give a good finish. Always wipe down the machine if used in the wet and make sure the chain is well oiled. Check that the control cables are kept well oiled (this will only take 5 minutes of your time each week with an oil can). There is also less chance of water getting into the cables, if they are well lubricated.

19 Fault diagnosis: frame and forks

Symptom	Cause	Remedy
Machine veers to left or right with hands off handlebars	Wheels out of alignment Forks twisted Frame bent	Check wheels and realign Strip and repair Strip and repair or renew
Machine tends to roll at low speeds	Steering head bearings not adjusted correctly or worn	Check adjustment and renew the bearings, if worn
Machine tends to wander	Worn swinging arm bearings	Check and renew bearings Check adjustment and renew
Forks judder when front brake is applied	Steering head bearings slack Forks worn on sliding surfaces	Dismantle, lubricate and adjust Dismantle and renew worn parts
Forks bottom	Short of oil	Replenish with correct viscosity oil
Fork action stiff	Fork legs out of alignment Bent shafts, or twisted yokes	Strip and renew, or slacken clamp bolts, front wheel spindle and top bolts. Pump forks several times, and tighten from bottom upwards
Machine tends to pitch badly	Defective rear suspension units, or ineffective fork damping	Check damping action Check the grade and quantity of oil in the front forks

Chapter 5 Wheels, brakes and tyres

For modifications, and information relating to later models, see Chapter 7

Contents

Specifications

	CB750K(Z)	CB750K LTD	CB750F	CB900F
Wheels				
Type	Wire spoked or Comstar	Comstar	Reversed Comstar	Comstar
Brakes				
Front	UK: Twin hydraulic disc US: Single hydraulic disc	Single hydraulic disc	Twin hydraulic disc	Twin hydraulic disc
Rear	Single leading shoe drum	Single leading shoe drum	Single hydraulic disc	Single hydraulic disc
Tyres				
Front	3.50H19 (4PR)	3.50H19 (4PR)	3.25H19 (4PR)	3.25V19 (4PR)
Rear	4.25H18 (4PR) (4.50H17 (4PR) 1980 on)	4.50H17 (4PR) (130/190-16, C model)	4.00H18 (4PR)	400V18 (4PR)
Tyre pressures (cold)				
Front, up to 90 kg (200 lb) load	28 psi (2.0 kg cm²)	28 psi (2.0 kg cm²)	28 psi (2.0 kg cm²)	32psi (2.25 kg cm²)
Front, above 90 kg (200 lb) load	28 psi (2.0 kg cm²)	28 psi (2.0 kg cm²)	28 psi (2.0 kg cm²)	32 psi (2.25 kg cm²)
Rear, up to 90 kg (200 lb) load	32 psi (2.25 kg cm²) [28 psi (2.0 kg cm²) with 450H17]	28 psi (2.0 kg cm²)	32 psi (2.25 kg cm²)	36 psi (2.50 kg cm²)
Rear, above 90 kg (200 lb) load	40 psi (2.80 kg cm²)	40 psi (2.80 kg cm²)	40 psi (2.80 kg cm²)	40 psi (2.80 kg cm²)

Torque settings

Component (all models)	kgf m	lbf ft
Wheel spindle clamp nuts	1.8 – 2.5	13 – 18
Wheel spindle nut (front)	5.5 – 6.5	40 – 47
Wheel spindle nut (rear)	8.0 – 10.0	58 – 72
Rear wheel sprocket	8.0 – 10.0	58 – 72
Rear brake torque arm	1.8 – 2.5	13 – 18
Brake disc	2.7 – 3.3	20 – 24
Caliper bracket	3.0 – 4.0	22 – 29

1 General description

The dohc Honda 750 and 900 models employ various wheel and brake combinations, according to the model and year of manufacture. Wheels are either of the conventional wire-spoked type or of one of two patterns of the Honda Comstar design. Front brakes consist of either single or twin hydraulic disc, whilst rear braking is by single disc or drum.

2 Front wheel: examination and renovation – wire spoked type

1   Place the machine on its centre stand so that the front wheel is clear of the ground. Spin the wheel by hand and check the rim for alignment, noting that the service limit is 2.0 mm (0.08 in). Small irregularities can be corrected by tightening the spokes in the affected area. Any flats in the wheel rim will be evident at the same time. In this latter case it will be necessary to have the wheel rebuilt with a new rim. The machine should not be run with a deformed wheel since this will have a very adverse effect on handling.

2   Check for loose or broken spokes. Tapping the spokes is a good guide to the correct tension; a loose spoke will always produce a different sound and should be tightened by turning the nipple in an anti-clockwise direction. Always check for run out by spinning the wheel again. If the spokes have to be tightened by an excessive amount, it is advisable to remove the tyre and tube as detailed in Section 23 of this Chapter. This will enable the protruding ends of the spokes to be ground off, thus preventing them from chafing the inner tube and causing punctures.

3 Front wheel: examination and renovation – Comstar type

1   The Comstar wheels differ in that they are built up with pressed spokes riveted to the hub and rim. As such, the wheel must be considered a single unit, as Honda do not offer any form of rebuilding facility. A number of private engineering shops offer this service, but it must be noted that Honda do not approve of this course of action.

2   Spin the wheel and check for rim alignment by placing a pointer close to the rim edge. If the total radial or axial alignment variation is greater than 2.00 mm (0.08 in) the manufacturers recommend that the wheel is renewed. This policy is, however, a counsel of perfection and in practice a larger runout may not affect the handling properties excessively. As remarked upon earlier, repair of a damaged wheel is not possible; the wheel must be renewed.

3   Check the rim for localised damage in the form of dents or cracks. The existence of even a small crack renders the wheel unfit for further use unless it is found that a permanent repair is possible using arc-welding. This method of repair is highly specialised and therefore the advice of a wheel repair specialist should be sought. Because tubeless tyres are used, dents may

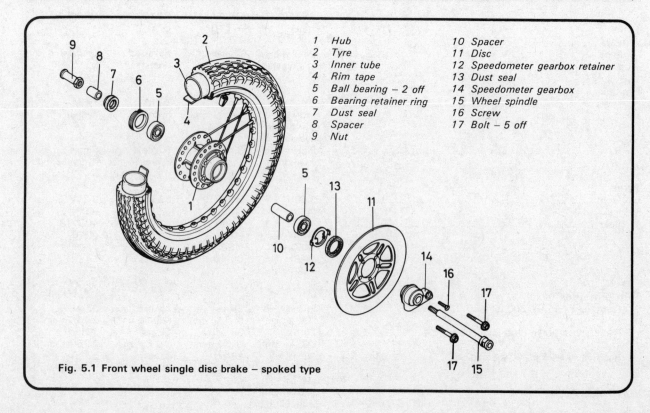

1	Hub	10	Spacer
2	Tyre	11	Disc
3	Inner tube	12	Speedometer gearbox retainer
4	Rim tape	13	Dust seal
5	Ball bearing – 2 off	14	Speedometer gearbox
6	Bearing retainer ring	15	Wheel spindle
7	Dust seal	16	Screw
8	Spacer	17	Bolt – 5 off
9	Nut		

Fig. 5.1 Front wheel single disc brake – spoked type

prevent complete sealing between the rim and tyre bead. This may not be immediately obvious until the tyre strikes a severely irregular surface, when the unsupported tyre wall may be deflected away from the rim, causing rapid deflation of the tyre. There again, specialist advice should be sought in order to establish whether continued use of the wheel is advisable.

4   Inspect the spoke blades for cracking and security. Check carefully the area immediately around the rivets which pass through the spokes and into the rim. In certain circumstances where steel spokes are fitted electrolytic corrosion may occur between the spokes, rivets and rim due to the use of different metals.

4   Front wheel: removal and replacement

1   With the front wheel supported well clear of the ground, remove the cross head screw which retains the speedometer cable to its drive gearbox, on the left-hand side of the hub. Pull the cable out and replace the screw, to prevent loss.

2   Remove the two bolts holding one of the caliper support brackets to the fork leg (twin disc models only) and lift the caliper and bracket assembly off the disc. Support the weight of the caliper with a length of string or wire attached to the frame or engine.

3   Slacken and remove the clamp nuts at the base of each fork leg. With the clamps released, the wheel will drop free and can be manoeuvred clear of the forks and mudguard.

4   Do not operate the front brake lever while the wheel is removed since fluid pressure may displace the pistons and cause leakage. Additionally, the distance between the pads will be reduced, making refitting of the brake discs more difficult.

5   Refit the wheel by reversing the dismantling procedure. Do not omit the spacer which is a push fit in the oil seal on the right-hand side of the wheel or the speedometer gearbox which is a push fit on the left-hand side. Ensure that the speedometer drive dogs engage with the notches in the gearbox drive sleeve. Lift the wheel into position and insert the wheel spindle so that

the spindle head is flush with the outer face of the clamp and fork leg. Tighten the clamp front nuts first to a torque setting of 1.8 – 2.5 kgf m (13 – 18 lbf ft). The rear nuts can now be tightened to the above setting, leaving a small gap at the **rear** of the clamp. Note that this is intentional, and thus the nuts must be secured in the above sequence. Refit the speedometer cable, and the brake caliper where appropriate.

6   Spin the wheel to ensure that it revolves freely and check the brake operation. Check that all nuts and bolts are fully tightened. If the clearance between the disc and pads is incorrect pump the front brake lever several times to adjust.

4.1 Remove screw and pull drive cable clear of wheel

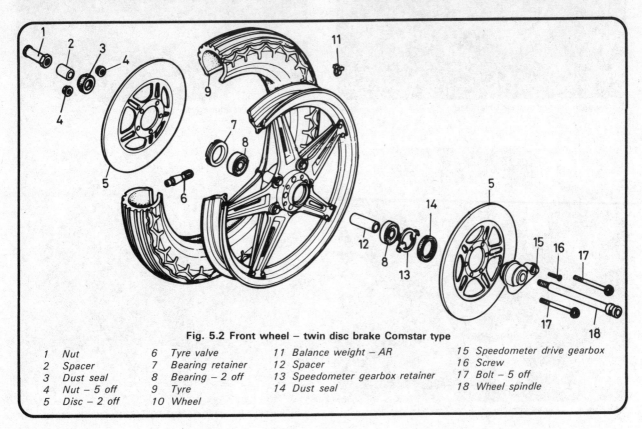

Fig. 5.2 Front wheel – twin disc brake Comstar type

1	Nut	6	Tyre valve	11	Balance weight – AR
2	Spacer	7	Bearing retainer	12	Spacer
3	Dust seal	8	Bearing – 2 off	13	Speedometer gearbox retainer
4	Nut – 5 off	9	Tyre	14	Dust seal
5	Disc – 2 off	10	Wheel		

15 Speedometer drive gearbox
16 Screw
17 Bolt – 5 off
18 Wheel spindle

4.2a Remove caliper mounting bolts ...

4.2b ... and move caliper clear of disc

4.3 Release clamps at base of fork legs

4.5a Note spacer during reassembly

4.5b Check that speedometer drive gearbox engages dogs

4.5c Arrows on spindle clamps must face forward

5 Front disc brake assembly: examination and brake pad renewal

1   Check the front brake master cylinder, hoses and caliper units for signs of leakage. Pay particular attention to the condition of the hoses, which should be renewed without question if there are signs of cracking, splitting or other exterior damage. Check the hydraulic fluid level by referring to the upper and lower level lines visible on the exterior of the translucent reservoir body.

2   Replenish the reservoir after removing the cap on the brake fluid reservoir and lifting out the diaphragm plate. The condition of the fluid is one of the maintenance tasks which should **never be neglected**. If the fluid is below the lower level mark, brake fluid of the correct specification must be added. **Never** use engine oil or any fluid other than that recommended. Other fluids have unsatisfactory characteristics and will rapidly destroy the seals.

3   The two sets of brake pads should be inspected for wear. Each has a red groove, which marks the wear limit of the friction material. When this limit is reached, both pads in the set must be renewed, even if only one has reached the wear mark. A small inspection window, closed by a plastic cap, is provided in the top of each caliper unit so that examination of pad condition may be carried out easily.

4   If the brake action becomes spongy, or if any part of the hydraulic system is dismantled (such as when a hose has been renewed) it is necessary to bleed the system in order to remove all traces of air. Follow the procedure in Section 8 of this Chapter.

5   The brake pads can be removed after the inspection cover has been removed, this being retained by a single screw. Release the wire clip which retains the two pad holding pins, withdrawing the latter to free the pads. The pads and shim can be pulled upwards.

6   Refit the new pads and replace the caliper by reversing the dismantling procedure. The caliper piston should be pushed inwards slightly so that there is sufficient clearance between the brake pads to allow the caliper to fit over the disc. Do not omit the anti-chatter shim which should be fitted on the rear face of the piston side pad with the arrow pointing forward. It is recommended that the outer periphery of the outer (piston) pad is lightly coated with disc brake assembly grease (silicone grease). Use the grease sparingly and ensure that grease **does not** come into contact with the friction surface of the pad.

5.1 Check that all unions and hoses are free from leaks

5.5a Remove screw and lift inspection cover clear

5.5b Displace ends of wire clip to free pins (arrowed)

5.5c Pull pins out to free pads

5.5d Pads can be pulled out for inspection or renewal

5.5e Note anti-squeal shim on backing plate

6 Front brake calipers: examination and overhaul

1   Where twin disc brakes are fitted, it is advisable that each
caliper is removed and overhauled separately, to prevent the
accidental transposition of identical components. Select a
suitable receptacle into which may be drained the hydraulic
fluid. Remove the banjo bolt holding the hydraulic hose at the
caliper and allow the fluid to drain. Take great care not to allow
hydraulic fluid to spill onto paintwork; it is a very effective paint
stripper. Hydraulic fluid will also damage rubber and plastic
components.
2   Remove the caliper from the fork leg and displace the brake
pads as described in the preceding Section. Withdraw the two
slider spindles and rubber boots from the support bracket.
3   Displace the circlip which holds the piston boot in position
and then prise out the piston boot, using a small screwdriver,
taking care not to scratch the surface of the cylinder bore. The
piston can be displaced most easily by applying an air jet to the
hydraulic fluid feed orifice. Be prepared to catch the piston as it
falls free. Displace the annular piston seal from the cylinder bore
groove.
4   Clean the caliper components thoroughly in trichlorethylene
or in hydraulic brake fluid. **CAUTION**: Never use petrol for
cleaning hydraulic brake parts otherwise the rubber components will be damaged. Discard all the rubber components as a
matter of course. The replacement cost is relatively small and
does not warrant re-use of components vital to safety. Check
the piston and caliper cylinder bore for scoring, rusting or
pitting. If any of these defects are evident it is unlikely that a
good fluid seal can be maintained and for this reason the
components should be renewed. Inspect the slider spindles for
wear and check their fit in the support bracket. Slack between
the spindles and bores may cause brake judder if wear is severe.
5   To assemble the caliper, reverse the removal procedure.
When assembling pay attention to the following points. Apply
caliper grease (high heat resistant) to the caliper spindles. Apply
a generous amount of brake fluid to the inner surface of the
cylinder and to the periphery of the piston, then reassemble. Do
not reassemble the piston with it inclined or twisted. When
installing the piston push it slowly into the cylinder while taking
care not to damage the piston seal. Apply brake pad grease
around the periphery of the moving pad. Bleed the brake after
refilling the reservoir with new hydraulic brake fluid, then check
for leakage while applying the brake lever tightly. Repeat the
entire procedure for the second brake caliper. After a test run,
check the pads and brake disc.
6   Note that any work on the hydraulic system must be
undertaken under ultra-clean conditions. Particles of dirt will
score the working parts and cause early failure.

7 Front disc brake master cylinder: examination and renovation

1   The master cylinder and hydraulic reservoir take the form of
a combined unit mounted on the right-hand side of the
handlebars, to which the front brake lever is attached. The
master cylinder is actuated by the front brake lever, and applies
hydraulic pressure through the system to operate the front
brake when the handlebar lever is manipulated. The master
cylinder pressurises the hydraulic fluid in the brake pipe which,
being incompressible, causes the piston to move in the caliper
unit and apply the friction pads to the brake disc. If the master
cylinder seals leak, hydraulic pressure will be lost and the
braking action rendered much less effective.
2   Before the master cylinder can be removed, the system
must be drained. Place a clean container below one caliper unit
and attach a plastic tube from the bleed screw on top of the
caliper unit to the container. Open the bleed screw one
complete turn and drain the system by operating the brake lever
until the master cylinder reservoir is empty. Close the bleed
screw and remove the pipe.
3   Remove the front brake stop lamp switch from the master
cylinder (where fitted). Unscrew the union bolt and disconnect
the connection between the brake hose and the master cylinder.
Unscrew the two master cylinder fastening bolts and remove
the master cylinder body from the handlebars. Empty any
surplus fluid from the reservoir.
4   Remove the brake lever from the body, remove the boot
stopper (taking care not to damage the boot) and then remove
the boot. Remove the circlip that was hidden by the boot, the
piston, primary cup, spring and check valve. Place the parts in
a clean container and wash them in new brake fluid. Examine
the cylinder bore and piston for scoring. Renew if scored. Check
also the brake lever for pivot wear, cracks or fractures, the hose
union threads and brake pipe threads for cracks or other signs
of deterioration.
5   When reassembling the master cylinder follow the removal
procedure in reverse order. Renew the various seals, lubricating
them with silicone grease or hydraulic fluid before they are
refitted. Make sure that the primary cup is fitted the correct way
round. Mount the master cylinder on the handlebars so that the
fluid reservoir is horizontal when the motorcycle is on the centre
stand with the steering in the straight ahead direction. Fill with
fresh fluid and bleed the system. Be sure to check the brake
reservoir by removing the reservoir cap. If the level is below the
ring mark inside the reservoir, refill to the level with the
prescribed brake fluid.
6   The component parts of the master cylinder assembly and
the caliper assemblies may wear or deteriorate in function over

a long period of use. It is however, generally difficult to foresee how long each component will work with proper efficiency. From a safety point of view it is best to change all the expendable parts every two years on a machine that has covered a normal mileage.

8 Bleeding the hydraulic brake system

1   If the hydraulic system has to be drained and refilled, if the front brake lever travel becomes excessive or if the lever operates with a soft or spongy feeling, the brakes must be bled to expel air from the system. The procedure for bleeding the hydraulic brake is best carried out by two persons.
2   First check the fluid level in the reservoir and top up with fresh fluid.
3   Keep the reservoir at least half full of fluid during the bleeding procedure.
4   Screw the cap on to the reservoir to prevent a spout of fluid

or the entry of dust into the system. Place a clean glass jar below one caliper bleed screw and attach a clear plastic pipe from the caliper bleed screw to the container. Place some clean hydraulic fluid in the jar so that the pipe is always immersed below the surface of the fluid.
5   Unscrew the bleed screw one half turn and squeeze the brake lever as far as it will go but do not release it until the bleeder valve is closed again. Repeat the operation a few times until no more air bubbles come from the plastic tube.
6   Keep topping up the reservoir with new fluid. When all the bubbles disappear, close the bleeder valve dust cap. Check the fluid level in the reservoir, after the bleeding operation has been completed. Repeat the air bleeding procedure with the second caliper unit (where appropriate).
7   Reinstall the diaphragm and tighten the reservoir cap securely. Do not use the brake fluid drained from the system, since it will contain minute air bubbles.
8   Never use any fluid other than that recommended. Oil must **not be used** under any circumstances.

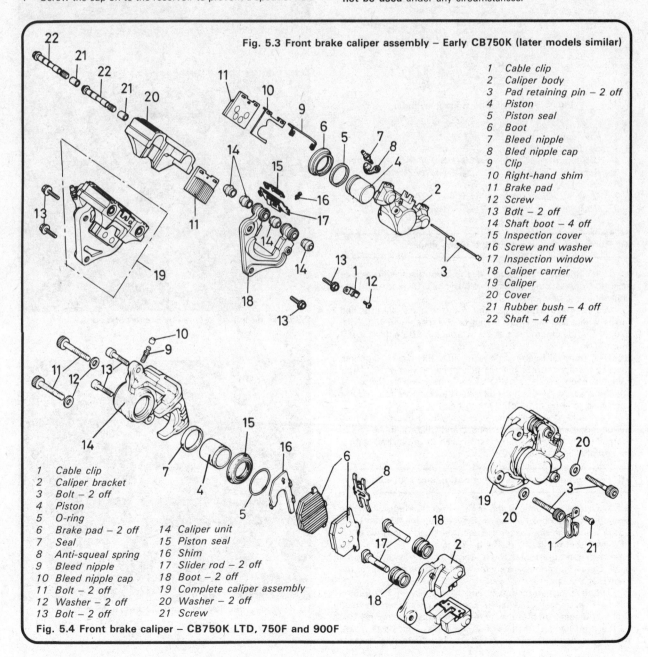

Fig. 5.3 Front brake caliper assembly – Early CB750K (later models similar)

1   Cable clip
2   Caliper body
3   Pad retaining pin – 2 off
4   Piston
5   Piston seal
6   Boot
7   Bleed nipple
8   Bled nipple cap
9   Clip
10 Right-hand shim
11 Brake pad
12 Screw
13 Bolt – 2 off
14 Shaft boot – 4 off
15 Inspection cover
16 Screw and washer
17 Inspection window
18 Caliper carrier
19 Caliper
20 Cover
21 Rubber bush – 4 off
22 Shaft – 4 off

1   Cable clip
2   Caliper bracket
3   Bolt – 2 off
4   Piston
5   O-ring
6   Brake pad – 2 off
7   Seal
8   Anti-squeal spring
9   Bleed nipple
10 Bleed nipple cap
11 Bolt – 2 off
12 Washer – 2 off
13 Bolt – 2 off
14 Caliper unit
15 Piston seal
16 Shim
17 Slider rod – 2 off
18 Boot – 2 off
19 Complete caliper assembly
20 Washer – 2 off
21 Screw

Fig. 5.4 Front brake caliper – CB750K LTD, 750F and 900F

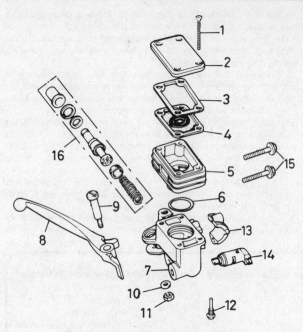

Fig. 5.5 Front brake master cylinder

1 Screw – 4 off	10 Washer
2 Filler cap	11 Nut
3 Diaphragm plate	12 Screw
4 Diaphragm	13 Handlebar fixing clamp
5 Reservoir	14 Front brake stop lamp
6 O-ring	switch
7 Master cylinder body	15 Bolt – 2 off
8 Brake lever	16 Piston assembly
9 Pivot bolt	

9 Removing and replacing the brake discs

1   It is unlikely that either of the two discs will require attention unless they become badly worn or scored or in the unlikely event of warpage.

2   Warpage should be measured with the discs still attached to the wheel and the wheel in situ on the machine using a dial gauge. The maximum permissible warpage for both discs is 0.3 mm (0.012 inch).

3   Disc removal is straightforward after the wheel has been taken out from the machine as described in Section 4 of this Chapter. Each disc is retained on the wheel hub by five bolts, which pass through the hub material. After removal of the bolts, each disc may be lifted off the hub boss.

4   The brake discs will wear eventually to a thickness which no longer provides sufficient support, and will probably begin to warp. The correct wear limit, which can be measured with a micrometer, is 6.0 mm (0.24 in).

10 Front wheel bearings: examination and replacement

1   Access is available to the front wheel bearings when the speedometer gearbox has been removed from the left-hand side of the wheel and the spacer has been removed from the centre of the oil seal on the right. The bearings are of the ball journal type and non-adjustable. There are two bearings and two oil seals, the two bearings are interposed by a distance collar in the centre of the hub.

2   The bearing arrangement and dismantling procedure is similar irrespective of whether spoked, Comstar or reversed Comstar wheels are fitted. Note that a threaded retainer is fitted to the right-hand side of the hub, and this must be removed before the bearings can be released. A suitable peg spanner

8.4 Use brake bleeder, or tube and glass jar, for brake bleeding operation

9.3 Brake disc(s) are secured by five through bolts

10.1 Remove wheel spindle and spacer to expose bearings

should be fabricated unless the Honda tool, No 07710-0010200, is available. Do not resort to using a punch to loosen the retainer; this will only result in damage. The retainer will be staked in position and will require firm, even pressure to release it. Note also that new bearings and seals should be fitted whenever the old items are removed, so check carefully for wear before dismantling commences.

3   The left-hand bearing should be drifted out first, from the right-hand side of the wheel. Use a long drift against the inner face of the inner race. It may be necessary to knock the collar to one side so that purchase can be made against the race. Work round in a circle to keep the bearing square in the housing. The oil seal and speedometer drive dog plate will be pushed out as the bearing is displaced. After removal of the bearing, take out the distance collar and then drift out the right-hand bearing and oil seal in a similar manner, from inside the hub.

4   Remove all the old grease from the hub and bearings, wash the bearings in petrol, and dry them thoroughly. Check the bearings for roughness by spinning them whilst holding the inner track with one hand and rotating the outer track with the other. If there is the slightest sign of roughness renew them.

5   Before driving the bearings back into the hub, pack the hub with new grease and also grease the bearings. Use a tubular drift to drive the bearings back into position. Do not omit to refit the oil seals and distance collar.

6   Fit the retainer into position, having obtained a new replacement if there were signs of wear or damage to the threads. Tighten the retainer firmly, then secure it in this position by staking at the junction of the retainer and the wheel hub.

11 Rear wheel: examination, removal and renovation

1   Place the machine on the centre stand so that the rear wheel is raised clear of the ground. Check for rim alignment, damage to the rim and loose or broken spokes by following the procedure relating to the front wheel, as described in Section 2 or 3 of this Chapter, depending on the wheel type fitted.

2   Detach the torque arm from the brake back plate after removing the nut secured by a split pin. Unscrew the brake rod adjuster nut fully and depress the brake pedal so that the rod leaves the trunnion in the brake operating arm. Refit the nut to secure the rod spring (drum brake models only).

3   Remove the split pin from the end of the wheel spindle and slacken the wheel spindle nut. Release the chain adjuster locknuts, then slacken off the adjusters to allow them to be pushed down through 90°. Displace the small fillets from the fork ends.

4   Push the wheel forwards to allow the final drive chain to be disengaged from the rear wheel sprocket. The wheel can now be pulled rearwards to clear the swinging arm, and then manoeuvred free of the rear of the machine. On rear disc brake models it is worth fitting a wooden wedge between the brake pads to prevent their expulsion should the brake pedal be operated whilst the wheel is removed.

5   Refit the wheel by reversing the dismantling procedure. Ensure that the torque arm is secure and that the securing split pins are fitted. Likewise do not omit the wheel spindle nut securing pin. Adjust the rear brake as described in Section 14 of this Chapter.

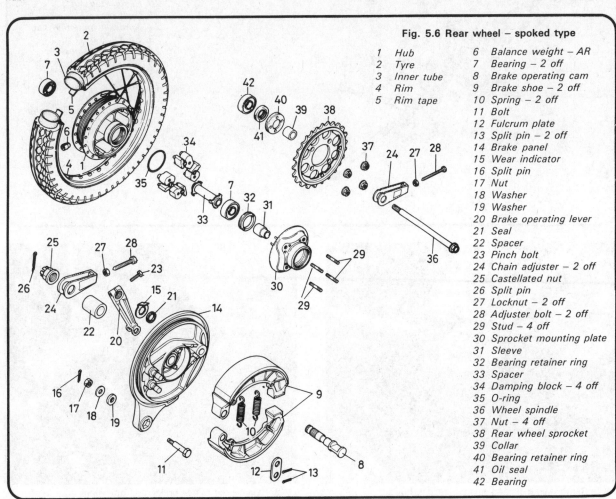

Fig. 5.6 Rear wheel – spoked type

1	Hub	6	Balance weight – AR
2	Tyre	7	Bearing – 2 off
3	Inner tube	8	Brake operating cam
4	Rim	9	Brake shoe – 2 off
5	Rim tape	10	Spring – 2 off
		11	Bolt
		12	Fulcrum plate
		13	Split pin – 2 off
		14	Brake panel
		15	Wear indicator
		16	Split pin
		17	Nut
		18	Washer
		19	Washer
		20	Brake operating lever
		21	Seal
		22	Spacer
		23	Pinch bolt
		24	Chain adjuster – 2 off
		25	Castellated nut
		26	Split pin
		27	Locknut – 2 off
		28	Adjuster bolt – 2 off
		29	Stud – 4 off
		30	Sprocket mounting plate
		31	Sleeve
		32	Bearing retainer ring
		33	Spacer
		34	Damping block – 4 off
		35	O-ring
		36	Wheel spindle
		37	Nut – 4 off
		38	Rear wheel sprocket
		39	Collar
		40	Bearing retainer ring
		41	Oil seal
		42	Bearing

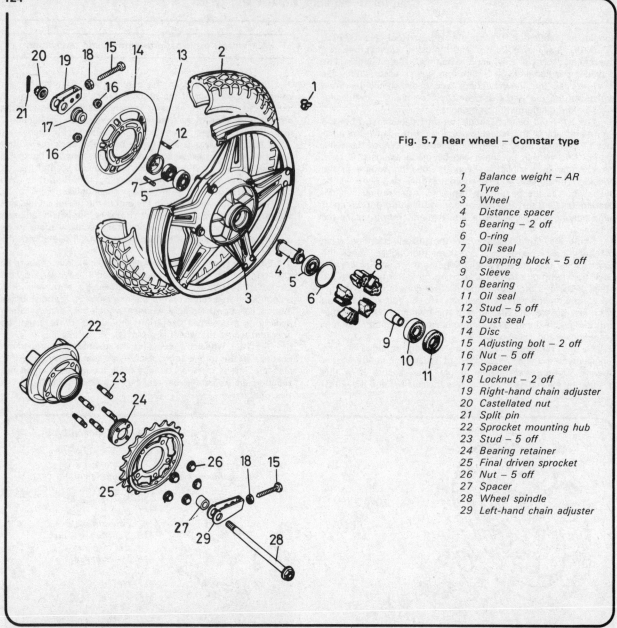

Fig. 5.7 Rear wheel – Comstar type

1 Balance weight – AR
2 Tyre
3 Wheel
4 Distance spacer
5 Bearing – 2 off
6 O-ring
7 Oil seal
8 Damping block – 5 off
9 Sleeve
10 Bearing
11 Oil seal
12 Stud – 5 off
13 Dust seal
14 Disc
15 Adjusting bolt – 2 off
16 Nut – 5 off
17 Spacer
18 Locknut – 2 off
19 Right-hand chain adjuster
20 Castellated nut
21 Split pin
22 Sprocket mounting hub
23 Stud – 5 off
24 Bearing retainer
25 Final driven sprocket
26 Nut – 5 off
27 Spacer
28 Wheel spindle
29 Left-hand chain adjuster

11.2a Remove split pin and nut to release torque arm

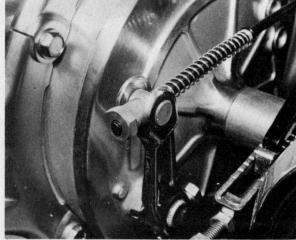

11.2b Release nut to free brake rod from wheel

11.3 Slacken wheel spindle, displace adjusters and remove fillets

11.4 Disengage drive chain and remove wheel

12 Rear wheel bearings: removal and replacement

1   Displace the sprocket carrier assembly by knocking it free of the cush drive rubbers in the hub. Note that both the sprocket carrier hub and the wheel hub feature a threaded retainer, similar to that fitted to the front hub (see Section 10). It follows that suitable peg spanner arrangements will be required unless the correct tools are available. The Honda part numbers for the latter are 07710-0010400 (retainer wrench body) 07710-0010100 (retainer wrench) and 07710-0010300 (retainer wrench). The accompanying photographs show the methods used for retainer removal without the appropriate tools.
2   Unscrew the two retainers to allow the bearings to be driven out in a similar manner to that described in Section 10. Examination and reassembly details are also similar to that given for the front wheel components.

13 Rear drum brake assembly: examination and renovation

1   On rear drum brake models, access to the rear brake components is gained after removing the rear wheel as described in Section 11. With the wheel removed, the brake backplate assembly can be lifted clear of the drum.
2   Note that the brake operating lever is marked with punch marks as a guide to reassembly. Check the position of these prior to removal, if this proves necessary. Note that it is advisable to dismantle, clean and lubricate the brake cam each time the brake is overhauled.
3   Examine the brake linings for oil, dirt or grease. Surface dirt can be removed with a stiff brush but oil soaked linings should be renewed. High spots can be carefully eased down with emery cloth.
4   Examine the condition of the brake linings and if they have worn to less than 2.0 mm (0.08 in) in thickness they should be renewed. The brake linings are bonded to the brake shoes and thus separate linings are not available.
5   To remove the shoes displace the split pin from each fulcrum post and lift off the link plate. Remove the pinch bolt from the operating arm and pull it off the splined shaft, followed by the wear indicator plate. Note the punch mark on the arm and shaft end to aid correct positioning on reassembly. Push the camshaft through from the outside whilst simultaneously easing the brake shoe ends off the fulcrum posts at the opposite side of the brake back plate. After removal, displace the camshaft from between the shoe ends and separate the shoes from the springs.
6   The brake drum should be checked for scoring. This

happens if the brake shoe linings have been allowed to get too thin. The drums should be quite smooth. Remove all traces of lining dust and wipe with a clean rag soaked in petrol to remove all traces of grease and oil.
7   Reassemble the brake back plate assembly by reversing the dismantling procedure. Check the return spring for wear or other damage at the hook ends and for stretching. Renew the springs, if necessary. Grease the operating camshaft and fulcrum posts with a high melting point grease before refitting the shoes. Note the O-ring on the camshaft, which prevents the escape of grease. When refitting the camshaft and arm, realign the marks to restore the original position.

14 Adjusting the rear drum brake

1   Adjustment of the rear brake is correct when there is 20 – 30 mm ($\frac{3}{4}$″ – 1″ approx) up and down movement measured at the rear brake pedal foot piece, between the fully off and on position. Adjustment is carried out by turning the nut on the brake rod.
2   The height of the brake pedal when at rest may be adjusted by means of the stop bolt which passes through a plate welded to the pedal shank. Loosen the locknut before making the adjustment and tighten it when adjustment is complete.
3   Either adjustment may require the brake pedal operated stop lamp switch to be re-adjusted.

12.1a Remove cush drive hub to reveal bearing retainer

12.1b Makeshift retainer wrench is crude but effective

12.1c Remove retainer to expose bearing

12.1d Bearings can be driven out of hub as shown

12.1e Note that sealed face of bearing must face outwards

12.1f Refit retainer and stake in position

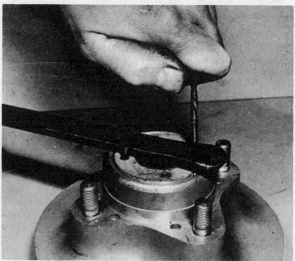

12.2a Cush drive bearing retainer can be removed as shown

12.2b Note seal fitted to inside of retainer

12.2c Retainer should be locked by staking

12.2d Do not omit plain spacer

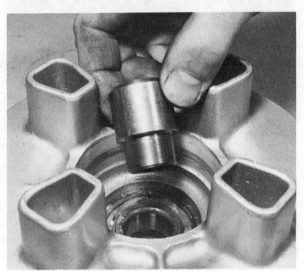

12.2e Stepped spacer is fitted from inside of bearing

13.1 Brake plate assembly can be removed as shown

13.4 Check brake linings for wear

15 Rear disc brake: examination and brake pad renewal

1   In the case of rear disc brake models, overhaul and maintenance follows a similar procedure to that described earlier for the front brake components. It follows that similar precautions must be taken to avoid the ingress of dirt, moisture or air into the system. The master cylinder is housed immediately behind the rear brake pedal and is connected by a short low-pressure hydraulic hose to the remote reservoir behind the right-hand side panel. A high-pressure hydraulic hose runs from the master cylinder to the rear brake caliper.

2   Check the unions and hose connections for signs of leakage, noting that prompt action must be taken if any such signs are noted. Bear in mind that although the hose between the reservoir and master cylinder is not under great pressure, any leakage could allow fluid spillage or even cause air to be admitted into the system.

3   Like the front brake pads, the rear pads carry a red line around the friction surface to denote the extent of allowable wear. These are visible through the small inspection window on the upper face of the caliper. To gain access to the pads for renewal it will be necessary to remove the caliper, leaving the caliper bracket assembly and pads in place. It is not necessary to disturb the hydraulic system.

4   Slacken the two Allen headed bolts which retain the caliper to the caliper bracket, and lift the caliper clear. Support the caliper to avoid placing undue strain on the hose or unions. The pads and their anti-squeal shim can now be removed in the same manner as the front caliper components.

5   When fitting new pads it should be noted that the arrow on the shim must face in the direction of normal wheel travel (towards the front of the machine). It will be necessary to depress the piston slightly when fitting the caliper over the new pads. This will allow sufficient clearance.

6   Note that the reservoir level should be checked as the fluid displaced from the caliper may take it above the normal maximum level. Check that the small bellows-type dust seals are in sound condition, because road dirt and water will cause rapid corrosion if it finds its way in. This applies equally to the main piston boot. If feasible, tighten the two Allen bolts to 1.5 – 2.0 kgf m (11 – 14 lbf ft).

16 Rear brake caliper: examination and overhaul

1   The procedure for overhauling the rear brake caliper is essentially the same as that described for the front brake in Section 6 of this Chapter. It should be noted that, in view of its location, the rear caliper is likely to be more seriously affected by the accumulation of road dirt than the front units. Appropriate steps should be taken to prevent the ingress of dirt into the caliper during removal.

2   Slacken the two Allen bolts and lift the caliper clear of the caliper bracket, but leave the hydraulic hose attached at this stage. Remove the bleed valve dust cap, then slacken the bleed valve, holding the caliper inverted over a drain tray. Operate the brake pedal repeatedly to expel the hydraulic fluid. Tighten the bleed valve and disconnect the hydraulic hose. Further dismantling of the caliper follows the dismantling sequence given in Section 6.

3   The caliper bracket supports the caliper above the rear disc, allowing a controlled sliding movement to equalise pressure between the pads when the brake is operated. If it is desired to remove the bracket from the machine, remove the torque arm bolt and displace the wheel spindle slightly to allow the bracket to be pulled clear.

4   Remove the caliper sliding pins and check for wear or corrosion. It is imperative that these move smoothly in the bracket if full braking effort is to be expected. Note that wear will allow the caliper to chatter during braking, necessitating renewal of the pins and/or bracket. It follows that the small bellows seals must function effectively if rapid wear is to be avoided. It is worth renewing these components during overhaul, as a precautionary measure. When reassembling the bracket, lubricate the pins with a medium weight high-temperature silicone grease.

17 Rear brake master cylinder: examination and renovation

1   Remove the right hand side panel to gain access to the hydraulic fluid reservoir. Slacken the hose clip which retains the low-pressure hydraulic feed pipe to the master cylinder. Cover the surrounding area with rags to protect the paintwork, then

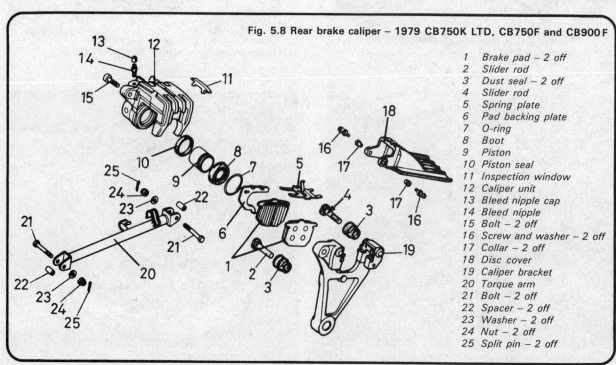

Fig. 5.8 Rear brake caliper – 1979 CB750K LTD, CB750F and CB900F

1   Brake pad – 2 off
2   Slider rod
3   Dust seal – 2 off
4   Slider rod
5   Spring plate
6   Pad backing plate
7   O-ring
8   Boot
9   Piston
10 Piston seal
11 Inspection window
12 Caliper unit
13 Bleed nipple cap
14 Bleed nipple
15 Bolt – 2 off
16 Screw and washer – 2 off
17 Collar – 2 off
18 Disc cover
19 Caliper bracket
20 Torque arm
21 Bolt – 2 off
22 Spacer – 2 off
23 Washer – 2 off
24 Nut – 2 off
25 Split pin – 2 off

pull off the pipe and allow the reservoir to drain into a suitable receptacle. Discard the fluid, which should not be re-used.

2   Working behind the alloy brake pedal plate, release the master cylinder pushrod from the brake pedal by withdrawing the split pin and displacing the clevis pin. The master cylinder can now be removed after releasing the two recessed Allen bolts which secure it to the alloy plate.

3   Pull back the rubber boot to expose the circlip which retains the pushrod assembly. Release the circlip and withdraw the pushrod. The master cylinder piston, seal and return spring can now be removed, noting that it may be necessary to displace them using compressed air or a footpump on the fluid outlet. Take care to avoid damage or injury through fluid splashes during this operation. Wrap the assembly in rag and look away whilst operating the air line or pump. For further details of the overhaul procedure, refer back to Section 7.

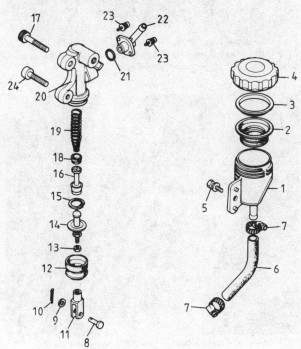

Fig. 5.9 Rear brake master cylinder and reservoir

1	Reservoir	14	Push rod
2	Diaphragm	15	Circlip
3	Diaphragm plate	16	Piston
4	Filler cap	17	Bolt
5	Bolt	18	Primary cup
6	Hose	19	Spring
7	Jubilee clip – 2 off	20	Master cylinder body
8	Clevis pin	21	O-ring
9	Washer	22	Union
10	Split pin	23	Screw and washer –
11	Clevis fork		2 off
12	Boot	24	Bolt
13	Nut		

18 Rear disc brake: bleeding the hydraulic system

1   As with the front brake, it will be necessary to bleed the rear hydraulic system whenever there has been a likelihood of air entering it. This will obviously apply if the system has been overhauled or if the caliper or master cylinder have been removed or dismantled. Follow the sequence described in Section 8, operating the rear brake pedal in place of the handlebar lever. Remember to keep the fluid level within limits during the bleeding operation.

19 Rear disc brake: checking the disc for wear

1   The rear brake disc should be examined for wear, noting that the likelihood of scoring is higher in the case of the rear item than for its front wheel counterpart, due to the location. Check for wear and warpage as described in Section 9.

20 Rear sprocket assembly: examination, renovation and replacement

1   The rear wheel sprocket is held to the wheel by four flanged nuts. The sprocket needs to be renewed only if the teeth are worn, hooked or chipped. It is always good policy to change both sprockets at the same time, also the chain, otherwise very rapid wear will develop.

2   It is not advisable to alter the rear wheel sprocket size or the gearbox sprocket size. The ratios selected by the manufacturer are the ones that give optimum performance with the existing engine power output.

21 Rear cush drive: examination and renovation

1   The cush drive assembly is contained in the left-hand side of the wheel hub. It takes the form of six triangular rubber pads incorporating slots. These engage with vanes on the coupling which is bolted to the rear sprocket. The rubbers engage with ribs on the hub and the whole assembly forms a shock absorber which permits the sprocket to move within certain limits. This cushions any surge or roughness in the transmission which would otherwise convey an impression of harshness.

2   The usual sign that shock absorber rubbers are worn is excessive movement in the sprocket, or rubber dust appearing in between the sprocket and hub. The rubbers should then be taken out and renewed.

22 Final drive chain: examination and lubrication

1   As the final drive chain is fully exposed on all models it requires lubrication and adjustment at regular intervals. To adjust the chain, take out the split pin from the rear wheel spindle and slacken the spindle nut. Undo the torque arm bolt, and leave the bolt in position, slacken the chain adjuster locknuts and turn the adjusters inwards to tighten the chain, or outwards to slacken the chain.

2   Chain tension is correct if there is 15 – 25 mm ($\frac{5}{8}$ – 1 inch) slack measured at the centre of the bottom run of the chain between the two sprockets. Note that excessive free play (2 inch/50 mm or more) will allow the chain to contact the frame, causing damage to both.

3   Do not run the chain too tight to try to compensate for wear, or it will absorb a surprising amount of engine power. Also it can damage the gearbox and rear wheel bearings.

4   All models are equipped with endless chains, having O-rings at the end of each pin to retain the grease used during assembly. It is not possible to remove the chain for full lubrication in Linklyfe or Chainguard as with normal chains. Lubrication is therefore restricted to frequent cleaning and lubrication with an aerosol chain lubricant of the type marked as suitable for use with O-ring chains.

5   Chain wear can be assessed by stretching the chain taut by means of the chain adjusters, and measuring a 20 link section. The measurement should be made between 21 pin centres. This length should normally be 317.5 mm (12.5 in). If worn to 323 mm (12.72 in) or more, the chain must be renewed.

6   Removal of the rear chain for renewal necessitates the removal of the rear wheel and swinging arm assembly, as described in Chapter 4, Section 9. Once the swinging arm has been removed the chain may be lifted off the gearbox sprocket. When fitting a new chain, ensure that the gearbox and rear wheel sprockets are in good condition.

20.1 Rear sprocket is retained by four nuts

21.1 Cush drive hub can be lifted clear of wheel and the rubbers checked for damage or compression

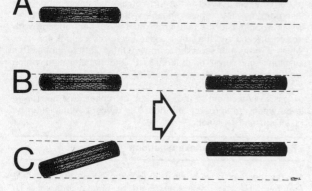

Fig. 5.10 Method of checking wheel alignment

A & C – Incorrect   B – Correct

22.1 Serrations on chain adjusters aid wheel alignment

23 Wheel balancing

1   It is customary on all high performance machines to balance the wheels complete with tyre and tube. The out of balance forces which exist are eliminated and the handling of the machine is improved in consequence. A wheel which is badly out of balance produces through the steering a most unpleasant hammering effect at high speeds.

2   Some tyres have a balance mark on the sidewall, usually in the form of a coloured spot. This mark must be in line with tyre valve, when the tyre is fitted to the inner tube. Even then the wheel may require the addition of balance weights, to offset the weight of the tyre valve itself.

3   If the wheel is raised clear of the ground and is spun, it will probably come to rest with the tyre valve or the heaviest part downward and will always come to rest in the same position. Balance weights must be added to a point diametrically opposite this heavy spot until the wheel will come to rest in ANY position after it is spun.

4   Balance weights are available in 20 gm (0.04 lb) and 30 gm (0.07 lb) sizes. Where a Comstar wheel is fitted the weight is attached to the wheel rim, but in the case of a wire spoked wheel the balance weight is fitted around the spoke nipple.

5   Note that because of the drag of the final drive chain the rear tyre must be balanced with the wheel removed from the machine, and supported on a suitable spindle.

24 Tubed tyres: removal and replacement – wire spoked wheel models

1   At some time or other the need will arise to remove and replace the tyres, either as a result of a puncture or because replacements are necessary to offset wear. To the inexperienced, tyre changing represents a formidable task, yet if a few simple rules are observed and the technique learned, the whole operation is surprisingly simple.

2   To remove the tyre from either wheel, first detach the wheel from the machine. Deflate the tyre by removing the valve core, and when the tyre is fully deflated, push the bead from the tyre away from the wheel rim on both sides so that the bead enters the centre well of the rim. Remove the locking ring and push the tyre valve into the tyre itself.

3   Insert a tyre lever close to the valve and lever the edge of the tyre over the outside of the rim. Very little force should be necessary; if resistance is encountered it is probably due to the fact that the tyre beads have not entered the well of the rim, all the way round.

4   Once the tyre has been edged over the wheel rim, it is easy to work round the wheel rim, so that the tyre is completely free from one side. At this stage the inner tube can be removed.

5 Now working from the other side of the wheel, ease the other edge of the tyre over the outside of the wheel rim that is furthest away. Continue to work around the rim until the tyre is completely free from the rim.

6 If a puncture has necessitated the removal of the tyre, reinflate the inner tube and immerse it in a bowl of water to trace the source of the leak. Mark the position of the leak, and deflate the tube. Dry the tube, and clean the area around the puncture with a petrol soaked rag. When the surface has dried, apply rubber solution and allow this to dry before removing the backing from the patch, and applying the patch to the surface.

7 It is best to use a patch of self vulcanizing type, which will form a very permanent repair. Note that it may be necessary to remove a protective covering from the top surface of the patch after it has sealed into position. Inner tubes made from a special synthetic rubber may require a special type of patch and adhesive, if a satisfactory bond is to be achieved.

8 Before replacing the tyre, check the inside to make sure that the article that caused the puncture is still not trapped inside the tyre. Check the outside of the tyre, particularly the tread area to make sure nothing is trapped that may cause a further puncture.

9 If the inner tube has been patched on a number of past occasions, or if there is a tear or large hole, it is preferable to discard it and fit a replacement. Sudden deflation may cause an accident, particularly if it occurs with the front wheel.

10 To replace the tyre, inflate the inner tube for it just to assume a circular shape but only to that amount, and then push the tube into the tyre so that it is enclosed completely. Lay the tyre on the wheel at an angle, and insert the valve through the rim tape and the hole in the wheel rim. Attach the locking ring on the first few threads, sufficient to hold the valve captive in its correct location.

11 Starting at the point furthest from the valve, push the tyre bead over the edge of the wheel rim until it is located in the central well. Continue to work around the tyre in this fashion until the whole of one side of the tyre is on the rim. It may be necessary to use a tyre lever during the final stages.

12 Make sure there is no pull on the tyre valve and again commencing with the area furthest from the valve, ease the other bead of the tyre over the edge of the rim. Finish with the area close to the valve, pushing the valve up into the tyre until the locking ring touches the rim. This will ensure that the inner tube is not trapped when the last section of bead is edged over the rim with a tyre lever.

13 Check that the inner tube is not trapped at any point. Reinflate the inner tube, and check that the tyre is seating correctly around the wheel rim. There should be a thin rib moulded around the wall of the tyre on both sides, which should be an equal distance from the wheel rim at all points. If the tyre is unevenly located on the rim, try bouncing the wheel when the tyre is at the recommended pressure. It is probable that one of the beads has not pulled clear of the centre well.

14 Always run the tyres at the recommended pressures and never under or over inflate. The correct pressures for solo use are given in the Specifications Section of this Chapter.

15 Tyre replacement is aided by dusting the side walls, particularly in the vicinity of the beads, with a liberal coating of french chalk. Washing up liquid can also be used to good effect, but this has the disadvantage of causing the inner surface of the wheel rim to rust.

16 Never replace the inner tube and tyre without the rim tape in position. If this precaution is overlooked there is a good chance of the ends of the spoke nipples chafing the inner tube and causing a crop of punctures.

17 Never fit a tyre that has a damaged tread or sidewalls. Apart from legal aspects, there is a very great risk of a blowout, which can have very serious consequences on a two wheeled vehicle.

18 Tyre valves rarely give trouble, but it is always advisable to check whether the valve itself is leaking before removing the tyre. Do not forget to fit the dust cap, which forms an effective extra seal.

25 Valves and valve caps: tubed tyres

1 Inspect the valves in the inner tubes from time to time making sure that the seal and spring are making an effective seal. There are tyre valve tools available for clearing damaged threads in the valve body, and incorporating thread clearing for the outside thread of the body. A key is also incorporated for tightening the valve core.

2 The valve caps prevent dirt and foreign matter from entering the valve, and also form an effective second seal so that in the event of the tyre valve sticking, air will not be lost.

3 Note that when a dust cap is fitted for the first time to a balanced wheel, the wheel may have to be rebalanced,

26 Tubeless tyres: removal and replacement – Comstar wheels

1 It is strongly recomended that should a repair to a tubeless tyre be necessary, the wheel is removed from the machine and taken to a tyre fitting specialist who is willing to do the job or taken to an official Honda dealer. This is because the force required to break the seal between the wheel rim and tyre bead is considerable and considered to be beyond the capabilities of an individual working with normal tyre removing tools. Any abortive attempt to break the rim to bead seal may also cause damage to the wheel rim, resulting in an expensive wheel replacement. If, however, a suitable bead releasing tool is available, and experience has already been gained in its use, tyre removal and refitting can be accomplished as follows.

2 To remove the tyre from either wheel, first detach the wheel from the machine by following the procedure in Sections 4 or 11 depending on whether the front or the rear wheel is involved. Deflate the tyre by removing the valve insert and when it is fully deflated, push the bead of the tyre away from the wheel rim on both sides so that the bead enters the centre well of the rim. As noted, this operation will almost certainly require the use of a bead releasing tool.

3 Insert a tyre lever close to the valve and lever the edge of the tyre over the outside of the wheel rim. Very little force should be necessary; if resistance is encountered it is probably due to the fact that the tyre beads have not entered the well of the wheel rim all the way round the tyre. Should the initial problem persist, lubrication of the tyre bead and the inside edge and lip of the rim will facilitate removal. Use a recommended lubricant, a dilute solution of washing-up liquid or french chalk. Lubrication is usually recommended as an aid to tyre fitting but its use is equally desirable during removal. The risk of lever damage to wheel rims can be minimised by the use of proprietary plastic rim protectors placed over the rim flange at the point where the tyre levers are inserted. Suitable rim projectors may be fabricated very easily from short lengths (4 – 6 inches) of thick-walled nylon petrol pipe which have been split down one side using a sharp knife. The use of rim protectors should be adopted whenever levers are used and, therefore, when the risk of damage is likely.

4 Once the tyre has been edged over the wheel rim. It is easy to work around the wheel rim so that the tyre is completely free on one side.

5 Working from the other side of the wheel, ease the other edge of the tyre over the outside of the wheel rim which is furthest away. Continue to work around the rim until the tyre is freed completely from the rim.

6 Refer to the following Section for details relating to puncture repair and the renewal of tyres. See also the remarks relating to the tyre valves in Section 27.

7 Refitting of the tyre is virtually a reversal of the removal procedure. If the tyre has a balance mark (usually a spot of coloured paint), as on the tyres fitted as original equipment, this must be positioned alongside the valve. Similarly, any arrow indicating direction of rotation must face the right way.

8 Starting at the point furthest from the valve, push the tyre

bead over the edge of the wheel rim until it is located in the central well. Continue to work around the tyre in this fashion until the whole of one side of the tyre is on the rim. It may be necessary to use a tyre lever during the final stages. Here again, the use of a lubricant will aid fitting. It is recommended strongly that when refitting the tyre only a recommended lubricant is used because such lubricants also have sealing properties. Do not be over generous in the application of lubricant or tyre creep may occur.

9   Fitting the upper bead is similar to fitting the lower bead. Start by pushing the bead over the rim and into the well at a point diametrically opposite the tyre valve. Continue working round the tyre, each side of the starting point, ensuring that the head opposite the working area is always in the well. Apply lubricant as necessary. Avoid using tyre levers unless absolutely essential, to held reduce damage to the soft wheel rim. The use of the levers should be required only when the final portion of bead is to be pushed over the rim.

10 Lubricate the tyre beads again prior to inflating the tyre, and check that the wheel rim is evenly positioned in relation to the tyre beads. Inflation of the tyre may well prove impossible without the use of a high pressure air hose. The tyre will retain air completely only when the beads are firmly against the rim edges at all points and it may be found when using a foot pump that air escapes at the same rate as it is pumped in. This problem may also be encountered when using an air hose, on new tyres which have been compressed in storage and by virtue of their profile hold the beads away from the rim edges. To overcome this difficulty, a torniquet may be placed around the circumference of the tyre, over the central area of the tread. The compression of the tread in this area will cause the beads to be pushed outwards in the desired direction. The type of torniquet most widely used consists of a length of hose closed at both ends with a suitable clasp fitted to enable both ends to be connected. An ordinary tyre valve is fitted at one end of the tube so that after the hose has been secured around the tyre it may be inflated, giving a constricting effect. Another possible method of seating beads to obtain initial inflation is to press the tyre into the angle between a wall and the floor. With the airline attached to the valve additional pressure is then applied to the tyre by the hand and shim, as shown in the accompanying illustration. The application of pressure at four points around the tyre's circumference whilst simultaneously applying the airhose will often effect an initial seal between the tyre beads and wheel rim, thus allowing inflation to occur.

11 Having successfully accomplished inflation, increase the pressure to 40 psi and check that the tyre is evenly disposed on the wheel rim. This may be judged by checking that the thin positioning line found on each tyre wall is equidistant from the rim around the total circumference of the tyre. If this is not the case, deflate the tyre, apply additional lubrication and reinflate. Minor adjustments to the tyre position may be made by bouncing the wheel on the ground.

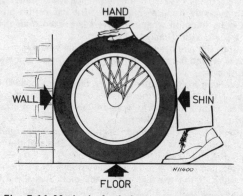

Fig. 5.11 Method of tubeless tyre inflation

Apply pressure at the four marked points to facilitate initial inflation

12 Always run the tyre at the recommended pressures and never under or over-inflate. The correct pressures for solo use are given in the Specification Section of this Chapter. If a pillion passenger is carried, increase the rear tyre pressure only as recommended.

27 Puncture repair and tyre renewal – tubeless tyre

1   The primary advantage of the tubeless tyre is its ability to accept penetration by sharp objects such as nails etc without loss of air. Even if loss of air is experienced, because there is no inner tube to rupture, in normal conditions a sudden blow-out is avoided.

2   If a puncture of the tyre occurs, the tyre should be removed for inspection for damage before any attempt is made at remedial action. The temporary repair of a punctured tyre by inserting a plug from the outside should not be attempted. Although this type of temporary repair is used widely on cars, the manufacturers strongly recommend that no such repair is carried out on a motorcycle tyre. Not only does the tyre have a thinner carcass, which does not give sufficient support to the plug, but the consequences of a sudden deflation are often sufficiently serious that the risk of such an occurrence should be avoided at all costs.

3   The tyre should be inspected both inside and out for damage to the carcass. Unfortunately the inner lining of the tyre – which takes the place of the inner tube – may easily obscure any damage and some experience is required in making a correct assessment of the tyre condition.

4   There are two main types of tyre repair which are considered safe for adoption in repairing tubeless motorcycle tyres. The first type of repair consists of inserting a mushroom-headed plug into the hole from the **inside** of the tyre. The hole is prepared for insertion of the plug by reaming and the application of an adhesive. The second repair is carried out by buffing the inner lining in the damaged area and applying a cold or vulcanised patch. Because both inspection and repair, if they are to be carried out safely, require experience in this type of work, it is recommended that the tyre be placed in the hands of a repairer with the necessary skills, rather than repaired in the home workshop.

5   In the event of an emergency, the only recommended 'get-you-home' repair is to fit a standard inner tube of the correct size. If this course of action is adopted, care should be taken to ensure that the cause of the puncture has been removed before the inner tube is fitted. It will be found that the valve hole in the rim is considerably larger than the diameter of the inner tube valve stem. To prevent the ingress of road dirt, and to help support the valve, a spacer should be fitted over the valve. A conversion spacer for most Honda models equipped with Comstar wheels is available from Honda dealers.

6   In the event of the unavailability of tubeless tyres, ordinary tubed tyres fitted with inner tubes of the correct size may be fitted. Refer to the manufacturer or a tyre fitting specialist to ensure that only a tyre and tube of equivalent type and suitability is fitted, and also to advise on the fitting of a valve nut/spacer to the rim hole.

28 Tyre valves: description and renewal – Comstar wheels

1   It will be appreciated from the preceding Sections that the adoption of tubeless tyres has made it necessary to modify the valve arrangement, as there is no longer an inner tube which can carry the valve core. The problem has been overcome by using a moulded rubber valve body which locates in the wheel rim hole. The valve body is pear-shaped, and has a groove around its widest point which engages with the rim forming an airtight seal.

2   The valve is fitted from the rim well, and it follows that it can only be renewed whilst the tyre itself is removed from the wheel. Once the valve has been fitted, it is almost impossible to remove it without damage, and so the simplest method is to cut

Deflate tyre. After releasing beads, push tyre bead into well of rim at point opposite valve. Insert lever adjacent to valve and work bead over edge of rim.

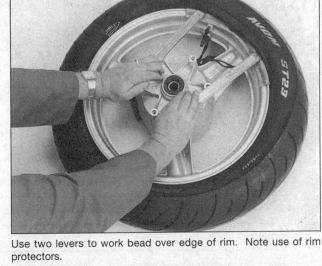

Use two levers to work bead over edge of rim. Note use of rim protectors.

When first bead is clear, remove tyre as shown.

Before fitting, ensure that tyre is suitable for wheel. Take note of any sidewall markings such as direction of rotation arrows.

Work first bead over the rim flange.

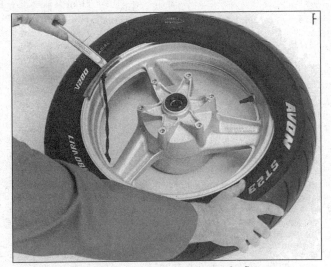

Use a tyre lever to work the second bead over rim flange.

it as close as possible to the rim well. The two halves of the old valve can then be removed.

3   The new valve is fitted by inserting the threaded end of the valve body through the rim hole, and pulling it through until the groove engages in the rim. In practice, a considerable amount of pressure is required to pull the valve into position, and most tyre fitters have a special tool which screws onto the valve end to enable purchase to be obtained. It is advantageous to apply a little tyre bead lubricant to the valve to ease its insertion. Check that the valve is seated evenly and securely.

4   The incidence of valve body failure is relatively small, and leakage only occurs when the rubber valve case ages and begins to perish. As a precautionary measure, it is advisable to fit a new valve when a new tyre is fitted. This will preclude any risk of the valve failing in service. When purchasing a new valve, it should be noted that a number of different types are available.

The correct type for use in the Comstar wheel is a Schrader 413, Bridgeport 193M or equivalent.

5   The valve core is of the same type as that used with tubed tyres, and screws into the valve body. The core can be removed with a small slotted tool which is normally incorporated in plunger type pressure gauges. Some valve dust caps incorporate a projection for removing valve cores. Although tubeless tyre valves seldom give trouble, it is possible for a leak to develop if a small particle of grit lodges on the sealing face. Occasionally, an elusive slow puncture can be traced to a leaking valve core, and this should be checked before a genuine puncture is suspected.

6   The valve dust caps are a significant part of the tyre valve assembly. Not only do they prevent the ingress of road dirt into the valve, but also act as a secondary seal which will reduce the risk of sudden deflation if a valve core should fail.

29 Fault diagnosis: wheels, brakes and tyres

Symptom	Cause	Remedy
Handlebars oscillate at low speed	Buckle or flat in wheel rim, most probably front wheel	Check rim for damage by spinning wheel. Renew wheel if not true
	Tyre pressure incorrect	Check, and if necessary adjust
	Tyre not straight on rim	Check tyre fitting. If necessary, deflate tyre and reposition
	Worn wheel or steering head bearings	Check and renew or adjust
Machine tends to weave	Tyre pressure incorrect	Check and if necessary adjust. If sudden, check for puncture
	Suspension worn or damaged	Check action of front forks and rear suspension units. Check swinging arm for wear.
Machine lacks power and accelerates poorly	Front or rear disc brake binding	Hot disc or caliper indicates binding. Overhaul caliper(s) and master cylinder, fit new pads if required, check disc(s) for scoring or warpage
	Rear drum brake binding	Hot drum indicates binding. Dismantle brake. Clean and lubricate pivots. Check springs. Check drum for scoring or warpage
Brakes grab or judder when applied gently	Rear drum brake: ends of linings not chamfered	File chamfer on leading edge of lining
	Rear disc brake: caliper bracket worn or loose; disc warped	Dismantle, clean and renew as required.
	Brake drum scored or warped	Remove wheel. Skim drum or renew
	Front brake: pads badly worn or scored. Wrong type of pad fitted	Renew pads and check disc and caliper
	Warped disc	Renew
Brake squeal		
Front and rear disc:	Glazed pads. Pads worn to backing metal	Sand pad surface to remove glaze, then use brake gently for about 100 miles to permit bedding in. If worn to backing check that disc is not damaged and renew as necessary
	Caliper and pads polluted with brake dust or foreign matter	Dismantle and clean. Overhaul caliper where necessary
Rear drum:	Accumulated brake dust in drum	Dismantle and clean
	Glazed linings	Restore as described for glazed pads
Excessive front brake lever travel	Air in system	Find cause of air's presence. If due to leak, rectify, then bleed brake
	Very badly worn pads	Renew, and overhaul system where required
	Badly polluted caliper	Dismantle and clean
Front brake lever feels springy	Air in system	See above
	Pads glazed	See above
	Caliper jamming	Dismantle and overhaul

Chapter 6 Electrical system

For modifications, and information relating to later models, see Chapter 7

Contents

Specifications

Battery

Make	Yuasa
Capacity	12 volt 14Ah
Polarity	Negative earth

Alternator

Type	Three-phase
Output	260W at 5000 rpm (18A min at 5000 rpm)

Voltage regulator/rectifier

Type	Integrated circuit, non-adjustable

Starter motor

Brush length	12.0 – 13.0 mm (0.47 – 0.51 in)
Service limit	7.5 mm (0.30 in)

Bulb wattage (all 12 volt)	CB750K(Z)	CB900F	CB750F	CB750K LTD
Headlamp:				
UK	60/55W H4 Quartz Halogen	60/55W H4 Quartz Halogen	60/55W Quartz Halogen	–
US	65/50W sealed beam unit	–	65/50W sealed beam unit	65/50W sealed beam unit
Parking lamp (UK only)	4W	4W	4W	4W
Stop/tail lamp:				
UK	5/21W x 2	5/21W x 2	5/21W x 2	–
US	8/27W x 2	–	8/27W x 2	8/27W
Registration plate lamp (UK only) ...	10W	–	–	–
Direction indicator lamps:				
UK	21W x 4	21W x 4	21W x 4	–
US	23W (rear x 2)	–	23W x 2 (rear)	23W x 2 (rear)
Position/indicator (US only) ...	8/23W (front x 2)	–	8/23W x 2 (front)	8/23W x 2 (front)
Indicator warning lamp	3.4W x 2	3.4W x 2	3.4W x 2	3.4W x 2
Oil pressure warning lamp	3.4W	3.4W	3.4W	3.4W
Neutral indicator lamp	3.4W	3.4W	3.4W	3.4W

High beam indicator lamp	3.4W	3.4W	3.4W	3.4W
Instrument illumination lamps	3.4W x 5	3.4W x 4	3.4W x 4	3.4W

Fuses

Main	30A	30A	30A	30A
1 (Neutral, oil)	10A	15A	10A	10A
2 (Headlamp)	10A	15A	10A	10A
3 (Indicators, brake, horn)	10A	15A	10A	10A
4 (Tail, instrument, position/ parking)	10A	15A	10A	10A

1 General description

All models are equipped with a twelve volt electrical system powered by a crankshaft-mounted alternator, or ac generator. An electromagnetic rotor is retained to the right-hand crankshaft end by means of a taper and central securing bolt. The stator and brush assemblies are retained inside the outer casing.

The resulting alternating current (ac) is fed to a combined regulator/rectifier unit mounted behind the left-hand side panel. This device converts the output to direct current (dc) and controls the system voltage. The regulated and rectified supply is then fed to the battery and electrical system.

2 Testing the electrical system: general

1 Checking the electrical output and the performance of the various components within the charging system requires the use of test equipment of the multi-meter type and also an ammeter of 0 - 0 ampere range. When carrying out checks, care must be taken to follow the procedures laid down and so prevent inadvertent incorrect connections or short circuits. Irreparable damage to individual components may result if reversal of current or shorting occurs. It is advised that unless some previous experience has been gained in auto-electrical testing the machine be returned to a Honda Service Agent or auto-electrician, who will be qualified to carry out the work and have the necessary test equipment.

2 If the performance of the charging system is suspect the system as a whole should be checked first, followed by testing of the individual components to isolate the fault. The three main components are the alternator, the rectifier and the regulator. Before commencing the tests, ensure that the battery is fully charged, as described in Section 6.

3 Charging system: checking the output

1 A quick check of the charging system condition may be made using the above-mentioned ammeter and voltmeter arrangement. Remove the side panels to gain access to the electrical components. Disconnect the red battery positive (+) lead, attaching it to the positive (+) terminal of the ammeter. Run a lead from the negative (-) ammeter terminal to the positive (+) battery terminal. Set the multimeter to the 0 - 20 volts dc scale (or higher) and connect the positive probe (+) probe to the positive (+) battery terminal. The negative (-) probe should be earthed.

2 Start the engine and allow it to reach normal running temperature, then turn on the main lights, set on main beam. At 1700 rpm, the discharge reading found at idle speed should be cancelled out. If the engine speed is now increased to 5000 rpm, the ammeter should show a zero reading or a slight charge, with 14 volts indicated on the voltmeter or multimeter.

3 If the output is erratic or noticeably below the specified amount, either the alternator or the rectifier may be at fault. The rectifier may be tested as described in Section 4. The alternator stator should be tested as follows, using a multi-meter set to the resistance function. Disconnect the block connector which connects the three alternator stator output leads plus the brush leads to the regulator/rectifier unit. The connector is located to the rear of the battery, behind the right-hand side panel. Using the multi-meter check that continuity exists between all three wires when tested in pairs. Check also that no lead has continuity with earth. If the results of the check do not correspond with those specified, there is evidence of short-circuits or open-circuits in the stator windings or the leads. The specified resistance is 0.41 − 0.51 ohms, though it will be sufficient just to note whether there is an open or short circuit.

4 Low output may also be caused by sticking or worn alternator brushes. These are located in a plastic holder on the inside of the alternator cover. Renew both brushes if either is worn down to or beyond the scribed wear limit line. The brushes are retained by two screws on the underside of the holder, which can be removed once the stator and brush holder securing screws have been removed and the assembly withdrawn from the cover.

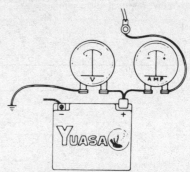

Fig. 6.1 Charging system output test connections

4 Regulator/rectifier unit: location and testing

1 If performance of the charging system is suspect, but the alternator is found to be in good condition, it is probable that one side of the combined regulator/rectifier unit is malfunctioning. Exactly which side is malfunctioning is of academic interest only because the sealed unit cannot be repaired, but must be renewed.

2 Voltage regulator performance may be checked using a voltmeter connected across the battery. Connect the positive voltmeter lead to the positive (+) battery terminal and the negative voltmeter lead to the negative (-) battery terminal. Start the engine and increase the speed until a 14 volt output is registered. Increase the engine speed to approximately 5,000 rpm. If the regulator is functioning correctly, the voltage will not rise to above 15 volts. If this voltage is exceeded, the unit is malfunctioning.

3 A further test of the unit's condition may be made by measuring the resistance of the rectifier circuit. To check these values accurately a meter or meters capable of reading in ohms and kilo ohms will be required. Failing this, most multimeters will be able to give a reliable indication despite reading solely on the latter scale. It is generally sufficient to distinguish between a very high resistance and a very low resistance when checking diode condition.

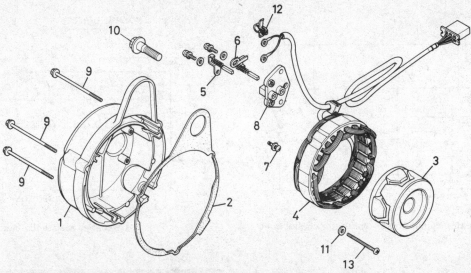

Fig. 6.2 Alternator assembly

1 Engine casing	5 Brush A	8 Brush retainer	11 Washer – 3 off
2 Gasket	6 Brush B	9 Bolt – 3 off	12 Cable clip
3 Rotor	7 Screw and washer – 2 off	10 Bolt	13 Screw – 3 off
4 Stator			

4   Before the test is started, it is helpful to understand what the rectifier is required to do. We have established that it converts ac to dc for charging purposes, and it does this by a matrix of diodes. Diodes, represented by a triangular symbol with a bar across one end in the accompanying diagram, can be imagined to act as one way valves, passing current in one direction only. Thus in our diagram the current should flow in the direction indicated by the point of the diode symbol, but must not flow back past the bar. In the event of rectifier failure, one or more of the diodes will have failed to operate correctly and thus will either pass current in both directions or will set up a high resistance in both directions. The purpose of the test is thus to check that the diodes are functioning correctly.

5   Trace the leads from the finned regulator/rectifier unit behind the left-hand side panel. These terminate in a 6-pin and a 4-pin connector block, each of which should be separated. Using an ohmmeter or a multimeter set on resistance, check the resistance between the green lead (4-pin connector) and each of the yellow leads in turn (6-pin connector). Reverse the meter probes and check the resistance between the green lead and each of the yellow leads in the opposite direction. If these diodes are in good condition there should be very little resistance (approx 5 – 40 ohms) in the normal direction of current flow and very high resistance (at least 2000 ohms) in the reverse direction.

6   Test the remaining diodes by connecting the meter between the red/white lead (4-pin connector) and each of the yellow leads (6-pin connector) in turn, noting the resistance reading obtained and then reversing the meter probes to obtain readings in the opposite direction. If these diodes are in good condition results should be the same as those given above. It follows that if in any of the tests the meter shows a very high or very low resistance in both directions then that diode is faulty and the complete regulator/rectifier unit must be renewed.

5   Battery: examination and maintenance

1   All models are fitted with a 12 volt battery of 14 ampere hour capacity.

2   The transparent plastic case of the battery permits the upper and lower levels of the electrolyte to be observed without disturbing the battery by removing the left-hand side cover. Maintenance is normally limited to keeping the electrolyte level between the prescribed upper and lower limits and making sure that the vent tube is not blocked. The lead plates and their separators are also visible through the transparent case, a further guide to the general condition of the battery.

3   Unless acid is spilt, as may occur if the machine falls over, the electrolyte should always be topped up with distilled water to restore the correct level. If acid is spilt onto any part of the machine, it should be neutralised with an alkali such as washing soda or baking powder and washed away with plenty of water, otherwise serious corrosion will occur. Top up with sulphuric acid of the correct specific gravity (1.260 to 1.280) only when spillage has occurred. Check that the vent pipe is well clear of the frame or any of the other cycle parts.

4   It is seldom practicable to repair a cracked battery case because the acid present in the joint will prevent the formation of an effective seal. It is always best to renew a cracked battery, especially in view of the corrosion which will be caused if the acid continues to leak.

5   If the machine is not used for a period, it is advisable to remove the battery and give it a refresher charge every six weeks or so from a charger. If the battery is permitted to discharge completely, the plates will sulphate and render the battery useless.

6   Occasionally, check the condition of the battery terminals to ensure that corrosion is not taking place and that the electrical connections are tight. If corrosion has occurred, it should be cleaned away by scraping with a knife and then using emery cloth to remove the final traces. Remake the electrical connections whilst the joint is still clean, then smear the assembly with petroleum jelly (NOT grease) to prevent recurrence of the corrosion. Badly corroded connections can have a high electrical resistance and may give the impression of a complete battery failure.

6   Battery: charging procedure

1   The normal charging rate for batteries of up to 14 amp hour capacity is $1\frac{1}{2}$ amps. It is permissible to charge at a more rapid rate in an emergency but this shortens the life of the battery, and should be avoided. Always remove the vent caps when recharging a battery, otherwise the gas created within the battery when charging takes place will explode and burst the case with disastrous consequences.

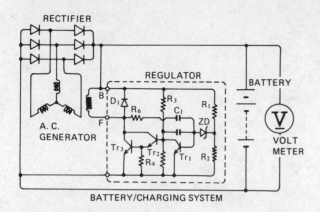

Fig. 6.3 Regulator/rectifier test – voltmeter connections

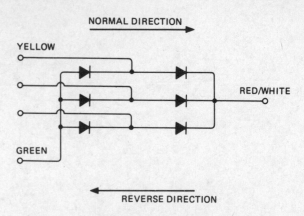

Fig. 6.4 Rectifier resistance check

5.2 Battery electrolyte level must be between two lines

5.6 Check that terminals are tight and corrosion-free

7   Fuse: location and replacement

1   The Honda dohc four cylinder models are equipped with a total of five fuses. The main fuse is housed in an enclosed fuse holder on the side of the starter solenoid. It is of the ribbon type and is rated at 30 amps. A spare fuse element is carried in the fuse holder. The remaining fuses are housed in a fuse holder immediately below the main instrument panel and can be reached after releasing the cover, this being screwed in place. The fuses, rated at 10 amps (15 amps in the case of the CB900F) protect the various electrical circuits, these being marked adjacent to each fuse. A spare fuse is housed in the fuse holder.

2   The fuses are incorporated in the system to give protection from a sudden overload as could happen with a short circuit. If a fuse blows it should not be renewed until the cause of the short is found. This will involve checking the electrical circuit to correct the fault. If this rule is not observed, the fuse will almost certainly blow again.

3   When a fuse blows and no spare is available, a get-you-home remedy is to wrap the fuse in silver paper before replacing it in the fuse holder. The silver paper will restore electrical continuity by bridging the broken wire within the fuse. Replace the doctored fuse at the earliest opportunity to restore full circuit protection. Make sure any short circuit is eliminated first.

4   Always carry spare fuses of the correct rating.

7.1a Main fuse (30 amp) is of the ribbon type

7.1b Secondary fuses are housed below instrument panel

8 Starter motor: removal, examination and replacement

1 An electric starter motor, operated from a small pushbutton on the right-hand side of the handlebars, provides the neccesary motive force for starting the engine. No kickstart lever is fitted. The starter motor is housed in a compartment to the rear of the cylinder block, beneath a light alloy cover.

2 When the starter button is operated, a heavy duty solenoid relay is activated, switching the necessary heavy current directly from the battery to the motor. The motor turns driving through an idler gear to the crankshaft-mounted roller clutch located on the left-hand side of the unit. The clutch works on the ramp and roller principle, ensuring that as soon as the engine starts the starter motor drive is disconnected automatically.

3 Before removing the motor, disconnect the positive (+) battery lead to isolate the electrical system. Release the starter motor cover by unscrewing the retaining bolts. Trace the heavy duty starter motor cable back to the solenoid and disconnect it after sliding back the protective boot. Release the starter motor mounting bolts, then remove the motor by drawing it to the right and then lifting it clear of its recess. The cable should be threaded clear of the frame.

4 The parts of the starter motor most likely to require attention are the brushes. The end cover is retained by the two long screws which pass through the lugs case on both end pieces. If the screws are withdrawn, the end cover can be lifted away and the brush gear exposed.

5 Lift up the spring clips which bear on the end of each brush and remove the brushes from their holders. The standard length and wear limit of the brushes is given in the Specifications at the beginning of this Chapter. If either brush is worn to a length less than that of the wear limit, renew the brushes as a pair.

6 Before the brushes are replaced, make sure the commutator on which they bear is clean. If necessary, the commutator segments may be burnished using Crokus paper. This is a fine abrasive paper produced for this purpose, and can be obtained from auto-electrical specialists. On no account use emery paper as it will damage the commutator. After cleaning, wipe the commutator with a rag soaked in methylated spirits or de-natured alcohol to remove any dust or grease.

7 Honda do not supply undercut figures for the mica insulators which lie netween each commutator segment, and thus imply that the armature assembly must be renewed when the commutator becomes worn. As this is a rather expensive solution, it may be worth seeking the advice of an auto-electrician who may be able to recondition the worn assembly at a favourable price.

8 Replace the brushes in their holders and check that they slide quite freely. Make sure the brushes are replaced in their original positions because they will have worn to the profile of the commutator. Replace and tighten the end cover, then replace the starter motor and cable in the housing, tighten down and re-make the electrical connection to the solenoid switch.

8.1 Starter motor is housed in crankcase recess

8.4 Release screws and remove starter motor end cover

8.5a Displace brush springs and pull brushes from holders

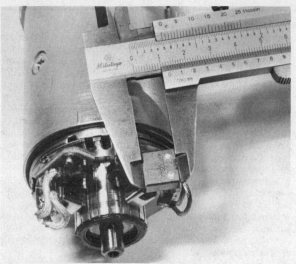

8.5b Check brush length as described in text

8.7a Check that commutator segments have not shorted using ohmmeter

8.7b Driving end of armature has shims ...

8.7c ... plus plain and tanged washers

8.7d Note oil seal in end cover

8.7e Ensure that brush plate locates correctly (arrowed)

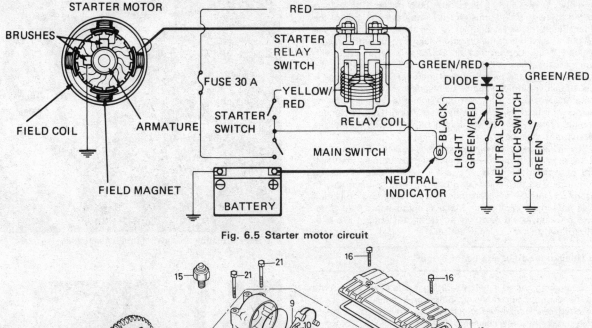

Fig. 6.5 Starter motor circuit

Fig. 6.6 Starter motor assembly

1	Starter idler gear	7	Cover	12	O-ring	17	Bolt – 3 off
2	Spindle	8	Starter motor assembly	13	Cover assembly	18	Roller – 3 off
3	Starter driven gear	9	Brush – 2 off	14	Seal	19	O-ring
4	Clutch hub	10	Spring – 2 off	15	Oil pressure switch	20	Washer – 2 off
5	Spring – 3 off	11	Bolt – 2 off	16	Screw – 2 off	21	Bolt – 2 off
6	Plunger – 3 off						

9 Starter solenoid switch: function and location

1 The starter motor switch is designed to work on the electro-magnetic principle. When the starter motor button is depressed, current from the battery passes through windings in the switch solenoid and generates an electro-magnetic force which causes a set of contact points to close. Immediately the points close, the starter motor is energised and a very heavy current is drawn from the battery.

2 This arrangement is used for two reasons. Firstly, the starter motor current is drawn only when the button is depressed and is cut off again when pressure on the button is released. This ensures minimum drainage on the battery. Secondly, if the battery is in a low state of charge, there will not be sufficient current to cause the solenoid contacts to close. In consequence it is not possible to place an excessive drain on the battery which in some circumstances, can cause the plates to overheat and shed their coatings. If the starter will not operate, first suspect a discharged battery. This can be checked by trying the horn or switching on the lights. If this check shows the battery to be in good shape, suspect the starter switch which should come into action with a pronounced click. It is located under the dualseat close to the battery, and can be identified by the heavy

duty starter cable connected to it. It is not possible to effect a satisfactory repair if the switch malfunctions; it must be renewed.

3 Before condemning the starter solenoid, check that the malfunction is not caused by a defective clutch interlock switch or neutral interlock switch. Unless either or both of these are working, ie neutral has been selected and/or the clutch is disengaged, the starter will not operate. Check the operation of both switches as described later in this Chapter.

10 Headlamp: replacement of bulbs

1 Depending on the model, the headlamp bulb is either held in a separate bulb holder, which fits into the rear of the reflector, or is an integral part of the reflector and therefore a sealed beam unit.

2 In both cases initial access is gained by removing the headlamp rim, complete with glass and reflector, which is retained by screws through the headlamp shell. The socket at the rear of the reflector may be pulled off the terminal pins.

3   On models fitted with a separate headlamp bulb, prise off the rubber boot which protects the bulb holder. The bulb is secured by two sprung arms which hinge from one side of the holder. Pinch the arms together to free them and then lift the bulb out. The bulb is of the tungsten-halogen type with a quartz envelope. It is important that the envelope is not touched with the fingers, because any greasy or acidic deposits will etch the quartz leading to the early failure of the bulb. If the envelope is touched inadvertently, it should be cleaned with a solvent such as methylated spirits. The pilot bulb holder is a press fit in the reflector, the bulb being of the bayonet fitting type. To remove the bulb, press it inwards and turn it anti-clockwise so that the bayonet pins disengage.

4   If a sealed beam headlamp becomes defective it is necessary to renew the complete glass/reflector unit as the bulb element is not detachable. The unit is held by screws passing through tabs on the headlamp rim. It will also be necessary to disconnect the beam adjustment screw and tension spring which pass into a captive nut. Make an approximate note of the screw position to aid resetting the beam adjustment. No pilot bulb is fitted to the sealed beam unit.

9.2 The starter solenoid switch/main fuseholder

---

11 Headlamp: adjusting beam height

---

1   Beam height is adjusted by rotating the headlamp about the two headlamp shell retaining bolts, after these have been slackened off. Horizontal adjustment is provided by a screw which passes into the headlamp rim.

2   UK lighting regulations stipulate that the lighting system must be arranged so that the light will not dazzle a person standing at a distance greater than 25 feet from the lamp, whose eye level is not less than 3 ft 6 inches above that plane. It is easy to approximate this setting by placing the machine 25 feet away from a wall, on a level road and setting the beam height so that it is concentrated at the same height as the distance of the centre of the headlamp from the ground. The rider must be seated normally during this operation and also the pillion passenger, if one is carried regularly.

10.2 Release headlamp unit by removing screws

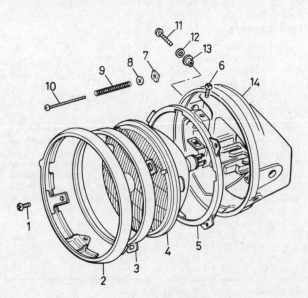

Fig. 6.7 Headlamp

1	Screw	
2	Rim	
3	Retaining ring	
4	Headlamp unit	
5	Mounting bracket	
6	Screw	
7	Nut	

8	Washer	
9	Spring	
10	Beam adjustment screw	
11	Screw – 2 off	
12	Collar – 2 off	
13	Insert – 2 off	
14	Headlamp shell	

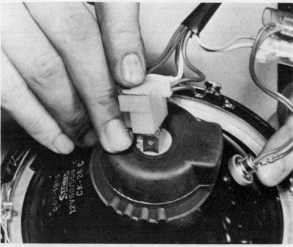

10.3a Remove connector and prise off rubber boot

10.3b Disengage clip by squeezing ends (arrowed)

10.3c Handle bulb thus – do not touch envelope

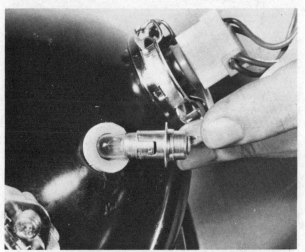

10.3d Pilot bulb (UK models) is push-fit in reflector

11.1 Headlamp adjustment is via screws in rim

12 Stop and tail lamp: replacing bulbs – KZ models

1 Two twin filament stop/tail lamp bulbs are used, each of which is fitted to a separate detachable holder. To gain access to the holders, the dualseat must be removed. Both holders pass through the rear wall of the seat fairing.
2 To remove a holder it should be twisted slightly so that the locating lugs disengage from the aperture, and then withdrawn. Both bulbs are of the bayonet type, with offset pins to prevent incorrect positioning.

13 Rear number plate lamp: replacing bulb – KZ models

1 Rear number plate illumination is provided by a separate 10 watt bulb which fits into a holder directly between and below the two stop/tail lamp bulb holders. The holder is a push fit in the seat fairing wall and is secured by a strip of metal held by a single screw.

14 Rear lamp assembly: removal and examination – KZ models

1 The rear lamp unit is unusually large and complicated on these machines, and forms part of the rear seat fairing. In the event of damage, a considerable amount of dismantling work is needed to remove the unit from the rear of the machine.
2 Start releasing the rear grab handle, followed by the tail fairing assembly, the latter being retained by cross head screws. The rear lamp bracket is now exposed, as are the three bolts which pass through it to retain the rear lamp assembly. Release the bolts and lift the unit away, having removed the three bulbholders as described in Sections 12 and 13.
3 The large lens can now be removed by unscrewing the four retaining screws which will have been exposed when the unit was removed from its mounting bracket. It will be noted that a double fresnel lens diffuser is fitted between the red outer lens and the bulbs. This focuses the light from the twin bulbs, and must be in place if the lamp assembly is to work properly. The diffuser is retained by three screws.

12.2a Stop/tail bulbs (KZ models) are fitted as shown

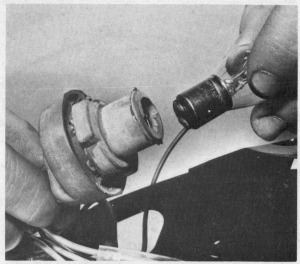

12.2b Bulbs are of the offset-pin bayonet type

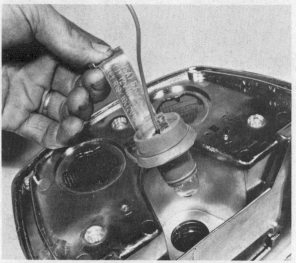

13.1 Separate number plate lamp is fitted (KZ models)

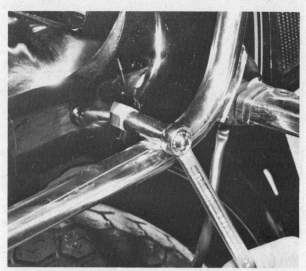

14.2a Release the grab rail ...

14.2b ... and tail fairing section (KZ models)

14.2c Remove the three securing bolts (arrowed) ...

14.2d ... and lift the lamp unit clear

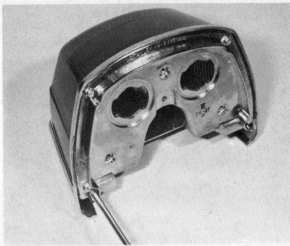

14.3a Remove the four lens screws ...

14.3b ... to release lens and expose diffuser

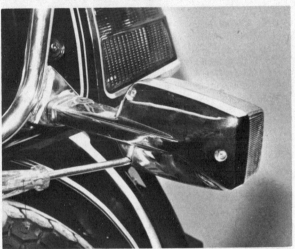

16.2a Indicator lens is retained by three screws

15 Rear lamp: bulb removal and replacement – FZ models

1 The rear lamp unit of the 750 and 900 F(Z) models differs in that it is a more conventional unit mounted beneath the tail of the seat fairing. Bulb access is by removing the lens, this being retained by four screws. The unit houses two twin-filament bulbs, there being no separate number (license) plate lamp.

16 Flashing indicator lamps: replacing bulbs

1 Flashing indicator lamps are fitted to the front and rear of the machine. They are mounted on short stalks through which the wires pass. Access to each bulb is gained by removing the two screws holding the plastic lens cover. All rear fitted bulbs are of the single-filament type as are the front bulbs of UK models. Machines supplied to the USA are fitted with double-filament bulbs with offset pins to prevent incorrect repositioning on replacement. The second filament is separately switched to allow the front indicators to serve as daylight running lamps.
2 To replace a bulb, remove the plastic lens cover by withdrawing the three retaining cross head screws. Push the bulb in, turn it to the left and withdraw. Note it is important to use a correctly rated bulb otherwise the flashing rate will be altered.

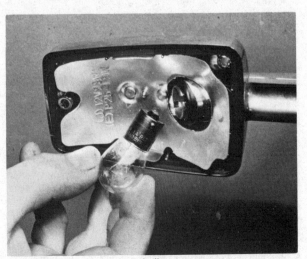

16.2b Release lens to expose bulb

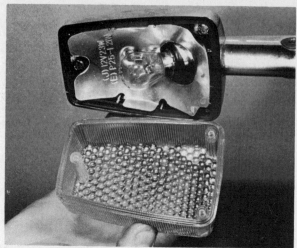

16.2c Bulb is bayonet type (twin filament – US model, front)

17 Flasher unit: location and replacement

1   The flasher relay unit is located behind the left-hand side panel and is supported on an anti-vibration mounting made of rubber.

2   If the flasher unit is functioning correctly, a series of audible clicks will be heard when the indicator lamps are in action. If the unit malfunctions and all the bulbs are in working order, the usual symptom is one initial flash before the unit goes dead; it will be necessary to replace the unit complete if the fault cannot be attributed to any other cause.

3   Care should be taken when handling the unit to ensure that it is not dropped or otherwise subjected to shocks. The internal components may be irreparably damaged by such treatment.

18 Instrument illumination and warning panel lamps: replacing the bulbs

1   To gain access to the instrument and warning panel bulbs it will be necessary to dismantle the instrument panel as described in Chapter 4 Section 4.12. This will allow access to the various instrument illumination and warning lamp bulbs. Check the recommended wattage with the Specification Section prior to renewal.

19 Horn: location and examination

1   The horn is suspended from a flexible steel strip bolted to the lower steering yoke behind the headlamp shroud.

2   A small screw and locknut arrangement provides adjustment. In the event that the horn's performance deteriorates significantly, experimentation with the screw setting will usually restore it to normal operation.

20 Ignition switch: removal and replacement

1   The combined ignition and lighting master switch is bolted to the upper yoke, and may be removed after the instrument panel has been detached. Disconnect the block connector plug from the base of the ignition switch. The switch is held in place by two screws, after the removal of which the switch can be displaced downwards.

2   Repair of a malfunctioning switch is not practicable as the component is a sealed unit; renewal is the only solution.

3   The switch may be refitted by reversing the dismantling sequence. Remember that a new switch will also require a new set of keys.

21 Stop lamp switch: adjustment

1   All models have a stop lamp switch fitted to operate in conjunction with the rear brake pedal. The switch is located immediately above the swinging arm on the left-hand side of the machine, it has a threaded body giving a range of adjustment.

2   If the stop lamp is late in operating, hold the body still and rotate the combined adjusting/mounting nut in an anti-clockwise direction so that the body moves away from the brake pedal shaft. If the switch operates too early or has a tendency to stick on, rotate the nut in a clockwise direction. As a guide, the light should operate after the brake pedal has been depressed by about 2 cm ($\frac{3}{4}$ inch). A stop lamp switch is also incorporated in the front brake system. The mechanical switch is a push fit in the handlebar lever stock. If the switch malfunctions, repair is impracticable. The switch should be renewed.

22 Handlebar switches: general

1   Generally speaking, the switches give little trouble but if necessary they can be dismantled by separating the halves which form a split clamp around the handlebars. Note that the machine cannot be started until the ignition cut-out on the right-hand end of the handlebars is turned to the central 'ON' position.

2   Always disconnect the battery before removing any of the switches, to prevent the possibility of a short circuit. Most troubles are caused by dirty contacts, but in the event of the breakage of some internal part, it will be necessary to renew the complete switch.

3   Because the internal components of each switch are very small, and therefore difficult to dismantle and reassemble, it is suggested a special electrical contact cleaner be used to clean corroded contacts. This can be sprayed into each switch, without the need for dismantling.

23 Neutral switch: general

1   A small switch fitted to the left-hand side of the crankcase operates a warning lamp in the instrument panel to indicate that neutral has been selected. More importantly, it is inter-connected with the starter solenoid and will only allow the engine to be started if the gearbox is in neutral, unless the clutch is disengaged. It can be checked by setting a multimeter on the resistance scale and connecting one probe to the switch terminal and the other to earth. The meter should indicate continuity when neutral is selected and infinite resistance when in any gear.

24 Clutch interlock switch: general

1   A small plunger-type switch is incorporated in the clutch lever, serving to prevent operation of the starter circuit when any gear has been selected, unless the clutch lever is held in. It can be checked by the method described above for the neutral switch. If defective it must be renewed, as there is no satisfactory means of repair. The switch can be removed after releasing the clutch cable and lever blade.

25 Switch testing procedure

1   In the event of a suspected malfunction the various switch contacts can be checked for continuity in a similar manner to that described in the two preceding Sections. Details of the switch connections and the appropriate wiring will be found in the colour wiring diagrams that follow. Note that the electrical system should be isolated by disconnecting the battery leads to avoid short circuits.

18.1 Instrument panel top removed to expose bulbs

19.2 Horn has adjustment screw to vary pitch

21.2 Stop lamp switch can be adjusted by turning body

22.1a Right-hand handlebar switch unit

22.1b Left-hand handlebar switch unit

23.1 Neutral switch screws into side of crankcase

26 Fault diagnosis: electrical system

Symptom	Cause	Remedy
Complete electrical failure	Blown fuse(s)	Check wiring and electrical components for short circuit before fitting new fuse
	Isolated battery	Check battery connections, also whether connections show signs of corrosion
Dim lights, horn and starter inoperative	Discharged battery	Remove battery and charge with battery charger. Check generator output and voltage regulator rectifier condition
Constantly blowing bulbs	Vibration or poor earth connection	Check security of bulb holders Check earth return connections
Parking lights dim rapidly	Battery will not hold charge	Renew battery at earliest opportunity
Tail lamp fails	Blown bulb or fuse	Renew
Headlamp fails	Blown bulb or fuse	Renew
Flashing indicators do not operate, or flash fast or slow	Blown bulb Damaged flasher unit	Renew bulb Renew flasher unit
Horn inoperative or weak	Faulty switch Incorrect adjustment	Check switch Adjust
Incorrect charging	Faulty alternator Faulty rectifier Faulty regulator Wiring fault	Check Check Check and adjust Check
Over or under charging	As above, or faulty battery	Check
Starter motor sluggish	Worn brushes Dirty commutator	Remove starter motor and renew brushes Clean
Starter motor does not turn	Machine in gear Emergency switch in OFF position Faulty switches or wiring Battery flat Loose battery terminal connection(s)	Disengage clutch Turn on Check continuity Recharge Check and tighten if necessary

The UK CB900 F2-D model

Engine/gearbox unit

Chapter 7 The 1981 to 1984 models

Contents

Specifications

The specifications shown below relate to the models covered in this update Chapter and appear in the sequence of the preceding Chapters

Model dimensions and weights – UK CB750 F-B, F2-C and F-D

Overall length	2240 mm (88.2 in)
Overall width	785 mm (30.9 in)
Overall height:	
CB750 F-B, F-D	1135 mm (44.7 in)
CB750 F2-C	1355 mm (53.3 in)
Wheelbase	1520 mm (59.8 in)
Seat height	815 mm (30.1 in)
Ground clearance	140 mm (5.5 in)
Dry weight:	
CB750 F-B, F-D	232 kg (511 lb)
CB750 F2-C	243 kg (536 lb)

Model dimensions and weights – UK CB900 F-B, F2-B, F-C, F2-C, F-D and F2-D

Overall length:	
CB900 F-B, F2-B	2240 mm (88.2 in)
CB900 F-C, F2-C, F-D, F2-D	2230 mm (87.8 in)
Overall width:	
CB900 F-B, F-C, F-D	805 mm (31.7 in)
CB900 F2-B, F2-C, F2-D	800 mm (31.5 in)
Overall height:	
CB900 F-B	1125 mm (44.3 in)
CB900 F2-B	1355 mm (53.3 in)
CB900 F-C, F-D	1115 mm (43.9 in)
CB900 F2-C, F2-D	1345 mm (53.0 in)
Wheelbase	515 mm (59.6 in)
Seat height	815 mm (30.1 in)
Ground clearance:	
CB900 F-B, F2-B	150 mm (5.9 in)
CB900 F-C, F2-C, F-D, F2-D	145 mm (5.7 in)
Dry weight:	
CB900 F-B	232 kg (511 lb)
CB900 F2-B	245 kg (540 lb)
CB900 F-C, F-D	242 kg (534 lb)
CB900 F2-C, F2-D	253 kg (558 in)

Model dimensions and weights – US CB750 K, F, C and SC 1981 to 83

Overall length:
CB750 K '81 '82	2295 mm (90.4 in)
CB750 F '81 '82	2195 mm (86.4 in)
CB750 C '81 '82	2300 mm (90.6 in)
CB750 SC '82 '83	2290 mm (90.2 in)

Overall width:
CB750 K '81 '82	890 mm (35.0 in)
CB750 F '81 '82	865 mm (34.1 in)
CB750 C '81 '82	920 mm (36.2 in)
CB750 SC '82 '83	850 mm (33.5 in)

Overall height:
CB750 K '81 '82	1155 mm (45.5 in)
CB750 F '81 '82	1150 mm (45.3 in)
CB750 C '81 '82	1165 mm (45.9 in)
CB750 SC '82 '83	1185 mm (46.7 in)

Wheelbase:
CB750 K '81 '82	1520 mm (59.8 in)
CB750 F '81 '82	1525 mm (60.0 in)
CB750 C '81 '82	1535 mm (60.4 in)
CB750 SC '82 '83	1545 mm (60.8 in)

Seat height:
CB750 K '81 '82	790 mm (31.1 in)
CB750 F '81 '82	810 mm (31.9 in)
CB750 C '81 '82	760 mm (29.9 in)
CB750 SC '82 '83	785 mm (30.9 in)

Footrest height:
CB750 K '81 '82	340 mm (13.4 in)
CB750 F '81 '82	350 mm (13.8 in)
CB750 C '81 '82	320 mm (12.6 in)
CB750 SC '82 '83	320 mm (12.6 in)

Ground clearance:
CB750 K '81 '82	145 mm (5.7 in)
CB750 F '81 '82	140 mm (5.5 in)
CB750 C '81 '82	130 mm (5.1 in)
CB750 SC '82 '83	Not available

Dry weight:
CB750 K '81 '82	234 kg (516 lb)
CB750 F '81 '82	230 kg (507 lb)
CB750 C '81 '82	234 kg (516 lb)
CB750 SC '82 '83	241 kg (531 lb)

Model dimensions and weights – US CB900 F 1981/82

Overall length	2195 mm (86.4 in)
Overall width	850 mm (33.5 in)
Overall height	1145 mm (45.1 in)
Wheelbase	1515 mm (59.6 in)
Seat height	815 mm (32.1 in)
Footrest height	350 mm (13.8 in)
Ground clearance	150 mm (5.9 in)
Dry weight	242 kg (532 lb)

Specifications UK CB750 models – except where shown all information relates to the F-B, F2-C and F-D models

Specifications relating to Chapter 2

Fuel tank
Total capacity	20 litres (4.4/5.3 Imp/US gal)
Reserve capacity	2.5 litres (0.55/0.66 Imp/US gal)

Carburettors
Type	VB52A

Specifications relating to Chapter 3

Spark plugs
Type:
Standard	NGK DR8ES or ND X27ESR-U
Cold climate (below 5°C)	NGK DR8ES-L or ND X24ESR-U

Specifications relating to Chapter 4

Front forks
Travel .... 160 mm (6.3 in)
Oil capacity (per leg) dry .... 170 cc
Oil capacity (per leg) at oil change .... 155 cc

Rear suspension
Swinging arm bearing type .... Needle roller

Torque settings

Component	kgf m	lbf ft
Rear wheel spindle	8.0 – 10.0	58 – 72
Swinging arm pivot nut	7.0 – 8.0	51 – 58
Rear brake torque arm	1.8 – 2.5	13 – 18
Rear suspension unit	3.0 – 4.0	22 – 29
Steering head adjuster nut	1.9 – 2.1	14 – 15

Specifications relating to Chapter 5

Brakes
Front:
CB750 F-B .... Twin hydraulic disc, single piston caliper
CB750 F2-C, F-D .... Twin hydraulic disc, twin piston caliper
Rear:
CB750 F-B .... Single hydraulic disc, single piston caliper
CB750 F2-C, F-D .... Single hydraulic disc, twin piston caliper

Tyre pressures (cold)

	rider only	rider and passenger
Front	32 psi (2.25 kg cm^2)	32 psi (2.25 kg cm^2)
Rear	32 psi (2.25 kg cm^2)	40 psi (2.80 kg cm^2)

Specifications relating to Chapter 6

Fuses
Main 30A
Neutral, oil 15A
Headlamp 15A
Turn signal, brake, horn 15A
Tail, instrument, position/parking 15A

Specifications UK CB900 F-B, F2-B, F-C, F2-C, F-D and F2-D models

Specifications relating to Chapter 1

Torque settings – F-B, F2-B

Component	kgf m	lbf ft
Crankcase bolts (10 mm)	5.5 – 6.0	40 – 43

Torque settings – F-C, F2-C, F-D, F2-D

Component	kgf m	lbf ft
Crankcase bolts (10 mm)	5.5 – 6.0	40 – 43
Automatic timing unit bolt	3.3 – 3.7	24 – 27
Engine mounting bolts:		
Engine mount to frame (10 mm)	3.5 – 4.5	25 – 33
Engine to front link (10 mm)	4.0 – 5.0	29 – 36
All 8 mm bolts	2.4 – 3.0	17 – 22
Right-hand frame section bolts	2.6 – 3.6	19 – 26

Specifications relating to Chapter 2

Fuel tank – all models
Total capacity .... 20 litres (4.4/5.3 Imp/US gal)
Reserve capacity .... 2.5 litres (0.55/0.66 Imp/US gal)

Carburettors – F-C, F2-C, F-D, F2-D
Pilot screw (turns out) .... $1\frac{7}{8}$

Specifications relating to Chapter 3

Torque wrench settings – F-C, F2-C, F-D, F2-D

Component	kgf m	lbf ft
Automatic timing unit bolt	3.3 – 3.7	24 – 27

Specifications relating to Chapter 4

Front forks

Type	Oil-damped telescopic, linked air assistance
Fork spring free length:	
CB900 F-B, F2-B	561.0 mm (22.09 in)
CB900 F-C, F2-C, F-D, F2-D	567.2 mm (22.33 in)
Service limit:	
CB900 F-B, F2-B	551.0 mm (21.69 in)
CB900 F-C, F2-C, F-D, F2-D	556.0 mm (21.89 in)
Fork lower leg ID:	
CB900 F-A	36.042 – 36.084 mm (1.4190 – 1.419 in)
CB900 F-B	38.040 – 38.080 mm (1.498 – 1.499 in)
CB900 F-C, F2-C, F-D, F2-D	40.040 – 40.080 mm (1.576 – 1.578 in)
Service limit:	
CB900 F-A	36.2 mm (1.425 in)
CB900 F-B, F2-B	38.2 mm (1.504 in)
CB900 F-C, F2-C, F-D, F2-D	40.2 mm (1.580 in)
Fork stanchion OD:	
CB900 F-A	34.925 – 34.950 mm (1.375 – 1.376 in)
CB900 F-B, F2-B	36.950 – 38.975 mm (1.455 – 1.456 in)
CB900 F-C, F2-C, F-D, F2-D	38.900 – 38.980 mm (1.531 – 1.535 in)
Service limit:	
CB900 F-A	34.85 mm (1.372 in)
CB900 F-B, F2-B	38.90 mm (1.531 in)
CB900 F-C, F2-C, F-D, F2-D	38.85 mm (1.530 in)
Bottom (stanchion) bush OD:	
CB900 F-A	35.94 – 36.00 mm (1.413 – 1.417 in)
CB900 F-B, F2-B	37.92 – 38.04 mm (1.493 – 1.498 in)
CB900 F-C, F2-C, F-D, F2-D	Not available
Service limit:	
CB900 F-A	35.85 mm (1.411 in)
CB900 F-B, F2-B	37.87 mm (1.539 in)
CB900 F-C, F2-C, F-D, F2-D	Not available
Top bush ID:	
CB900 F-A	35.07 – 35.13 mm (1.381 – 1.383 in)
CB900 F-B, F2-B	38.97 – 39.04 mm (1.534 – 1.537 in)
CB900 F-C, F2-C, F-D, F2-D	Not available
Service limit:	
CB900 F-A	35.25 mm (1.388 in)
CB900 F-B, F2-B	39.09 mm (1.539 in)
CB900 F-C, F2-C, F-D, F2-D	Not available
Fork air pressure:	
CB900 F-A	0 psi
CB900 F-B, F2-B, F-C, F2-C, F-D, F2-D	11 – 14 psi (0.8 – 1.0 kg/cm^2)
Maximum air pressure	See Section 5
Oil capacity (per leg) dry:	
CB900 F-A	190 ± 2.5 cc (6.6 ± 0.08 fl oz)
CB900 F-B, F2-B	320 cc (11.26 fl oz)
CB900 F-C, F2-C, F-D, F2-D	395 ± 2.5 cc (13.9 ± 0.08 fl oz)
Oil capacity (per leg) at oil change:	
CB900 F-A	170 ± 2.5 cc (5.9 ± 0.08 fl oz)
CB900 F-B, F2-B	300 cc (10.56 fl oz)
CB900 F-C, F2-C, F-D, F2-D	375 ± 2.5 cc (13.2 ± 0.08 fl oz)

Anti-dive components – F-C, F2-C, F-D, F2-D

Piston OD	17.947 – 17.980 mm (0.7066 – 0.7079 in)
Service limit	17.93 mm (0.716 in)
Spring free length	28.80 mm (1.134 in)
Service limit	28.2 mm (1.11 in)

Rear suspension – all models

Swinging arm bearing type	Needle roller

Torque settings – all models unless shown otherwise

Component	kgf m	lbf ft
Front fork air valve	0.4 – 0.7	3 – 5
Front fork cap bolt	1.5 – 3.0	11 – 22
Damper rod Allen bolt	1.5 – 2.5	11 – 18
Air hose union:		
Left	1.5 – 2.0	11 – 15
Right	0.4 – 0.7	3 – 5
Rear wheel spindle nut F-C, F2-C, F-D, F2-D	8.0 – 10.0	58 – 72

Swinging arm pivot nut F-C, F2-C, F-D, F2-D ....   6.0 – 8.0   43 – 59
Steering head adjuster nut F-C, F2-C, F-D, F2-D ....   1.8 – 2.0   13 – 14

Specifications relating to Chapter 5

Brakes – all models

Front .... Twin hydraulic disc, twin piston caliper
Rear .... Single hydraulic disc, twin piston caliper

Tyre pressures (cold) – F-C, F2-C, F-D, F2-D

	Up to 90 kg (200 lb) load	Above 90 kg (200 lb) load
Front	36 psi (2.50 kg cm²)	36 psi (2.50 kg cm²)
Rear	36 psi (2.50 kg cm²)	42 psi (2.90 kg cm²)

Torque settings – all models unless shown otherwise

Component	kgf m	lbf ft
Caliper shaft	2.5 – 3.0	18 – 22
Caliper mounting bolt – F-B, F2-B	2.2 – 2.5	16 – 18
Caliper mounting bolt – F-C, F2-C, F-D, F2-D:		
Upper	3.5 – 4.5	25 – 33
Lower	2.0 – 2.5	14 – 18
Caliper to bracket bolt – F-C, F2-C, F-D, F2-D	2.0 – 2.5	14 – 18
Pad pin retainer bolt	0.8 – 1.3	6 – 9
Anti-dive unit mounting bolt	0.6 – 0.9	4.3 – 6.5
Rear wheel spindle nut – F-C, F2-C, F-D, F2-D	8.0 – 10.0	58 – 72

Specifications relating to Chapter 6

Fuses – all models

Main 30A
Neutral, oil 15A
Headlamp 15A
Turn signal, brake, horn 15A
Tail, instrument, position/parking 15A

Specifications – US CB750 models. Unless stated, information applies to all K, F, C 1981 to 1982 and SC 1982 to 1983 models

Specifications relating to Chapter 1

Gearbox

Final reduction:
CB750 K, F 2.555:1
CB750 C, SC 2.388:1

Specifications relating to Chapter 2

Fuel tank

Total capacity:
CB750 K, F .... 20 litres (4.4/5.3 Imp/US gal)
CB750 C .... 16.5 litres (3.6/4.4 Imp/US gal)
Reserve capacity:
CB750 K .... 3.0 litres (0.66/0.80 Imp/US gal)
CB750 F .... 2.5 litres (0.55/0.66 Imp/US gal)
CB750 C .... 2.8 litres (0.62/0.70 Imp/US gal)

Carburettors

Type VB42A

Specifications relating to Chapter 3

Spark plugs

	Standard type		Optional type	
Make	NGK	ND	NGK	ND
Grade – standard fitment:				
CB750, all models (81)	D8EA	X24ES-U	DR8ES-L	X24ESR-U
CB750 K, F, C (82)	DR8ES-L	X24ESR-U	Not available	
CB750 SC, (82-83)	DR8ES-L	X22ESR-U	Not available	

Grade – cold climate (below 5°C, 41°F):				
CB750, all models (81)	D7EA	X22ES-U	DR7ES	X22ESR-U
CB750 K, F, C (82)	DR7ES	X22ESR-U	Not available	
CB750 SC, (82-83)	DR7ES	X22ESR-U	Not available	
Grade – Extended high speed:				
CB750, all models (81)	D9EA	X27ES-U	DR8ES	X27ESR-U
CB750 K, F, C (82)	DR8ES	X27ESR-U	Not available	
CB750 SC, (82-83)	DR8ES	X27ESR-U	Not available	

Specifications relating to Chapter 4

Front forks

Type:
- CB750 K, C, F (81, 82) .... Hydraulically damped coil spring telescopic, linked air assistance
- CB750 SC (82, 83) .... Hydraulically damped coil spring telescopic, linked air assistance TRAC anti-dive

Travel .... 160 mm (6.3 in)

Oil capacity (per leg) dry:
- CB750 K, (81, 82) .... 210 cc (7.1 fl oz)
- CB750 C, F (81, 82) .... 245 cc (8.2 fl oz)

Oil capacity (per leg) at oil change:
- CB750 K (81, 82) .... 190 cc (6.5 fl oz)
- CB750 C, F (81, 82) .... 225 cc (7.6 fl oz)

Oil capacity CB750 SC (82, 83):
- Left leg .... 357.5 – 362.5 cc (12.09 – 12.26 fl oz)
- Right leg .... 347.5 – 352.5 cc (11.75 – 11.92 fl oz)

Fork spring free length:
- CB750 K, C (81, 82) .... 551.0 mm (21.69 in)
- CB750 F (81, 82) .... 503.7 mm (19.80 in)
- CB750 SC (82, 83) .... 571.4 mm (22.50 in)

Service limit:
- CB750 K, C (81, 82) .... 541.0 mm (21.30 in)
- CB750 F (81, 82) .... 489.0 mm (19.30 in)
- CB750 SC (82, 83) .... 560.0 mm (22.05 in)

Fork stanchion OD:
- CB750 K, C (81, 82) .... 34.975 – 34.950 mm (1.377 – 1.376 in)
- CB750 F (81, 82) .... 36.950 – 36.975 mm (1.455 – 1.456 in)
- CB750 SC (82, 83) .... Not available

Service limit:
- CB750 K, C (81, 82) .... 34.85 mm (1.372 in)
- CB750 F (81, 82) .... 36.90 mm (1.453 in)
- CB750 SC (82, 83) .... Not available

Fork stanchion max, runout:
- CB750 except SC (81, 82) .... 0.2 mm (0.01 in)
- CB750 SC (82, 83) .... Not available

Fork air pressure:
- CB750 K, C, SC (81, 82, 83) .... 0.7 – 1.1 kg cm² (10 – 16 psi)
- CB750 F (81, 82) .... 0.8 – 1.2 kg cm² (11 – 17 psi)

Rear suspension

Swinging arm bearing type .... Needle roller

Torque settings – all models unless shown otherwise

Component	kgf m	lbf ft
Front fork cap bolt	1.5 – 3.0	11 – 22
Front fork damper rod Allen bolt –		
CB750 SC (82-83)	1.5 – 2.5	11 – 18
Front wheel spindle pinch bolt – CB750 C (81, 82)	1.5 – 2.5	11 – 18
Rear wheel spindle nut	8.0 – 10.0	58 – 72
Swinging arm pivot nut	7.0 – 8.0	51 – 58
Rear brake torque arm:		
CB750 K, C (81, 82) front	1.8 – 2.5	13 – 18
CB750 K, C (81, 82) rear	0.8 – 1.2	6 – 9
CB750 F (81, 82)	1.8 – 2.5	13 – 18
Rear suspension unit – upper and lower	3.0 – 4.0	22 – 29
Steering head adjuster nut	1.9 – 2.1	14 – 15
Fork air hose connections:		
Hose union to fork cap bolt	0.4 – 0.7	3 – 5
Connector to fork cap bolt	0.4 – 0.7	3 – 5
Hose union to connector	1.5 – 2.0	40 – 47

Specifications relating to Chapter 5

Brakes

Front:
CB750 K (81)	Single hydraulic disc, single piston caliper
CB750 K (82)	Single hydraulic disc, twin piston caliper
CB750 C (81)	Twin hydraulic disc, single piston caliper
CB750 C (82)	Twin hydraulic disc, twin piston caliper
CB750 F (81, 82)	Twin hydraulic disc, twin piston caliper
CB750 SC (82-83)	Twin hydraulic disc, twin piston caliper

Rear:
CB750 K, C (81, 82)	Single leading shoe drum
CB750 F (81, 82)	Single hydraulic disc, twin piston caliper
CB750 SC (82, 83)	Single leading shoe drum

Disc thickness (CB750 F 81-82, C '82, SC '82-83):
Front	4.9 – 5.1 mm (0.19 – 0.20 in)
Service limit	4.0 mm (0.16 in)
Rear – F only	6.9 – 7.1 mm (0.27 – 0.28 in)
Service limit	6.0 mm (0.24 in)

Disc thickness – CB750 K '82:
Front	6.9 – 7.1 mm (0.27 – 0.28 in)
Service limit	6.0 mm (0.24 in)

Disc maximum runout – CB750 all models '81 – front and rear ... 0.3 mm (0.012 in)

Master cylinder bore ID – CB750 F, '81-82, C, '82, SC '82-83:
Front	15.870 – 15.913 mm (0.6248 – 0.6255 in)
Service limit	15.925 mm (0.6270 in)
Rear – F only	14.000 – 14.043 mm (0.6248 – 0.6265 in)
Service limit	14.055 mm (0.5533 in)

Master cylinder bore ID – CB750 K '82:
Front	14.000 – 14.043 mm (0.6248 – 0.6265 in)
Service limit	14.055 mm (0.5533 in)

Master cylinder piston OD – CB750 F, '81-82, C '82 SC '82-83:
Front	15.827 – 15.854 mm (0.6231 – 0.6242 in)
Service limit	15.815 mm (0.6226 in)
Rear – F only	13.957 – 13.984 mm (0.5495 – 0.5506 in)
Service limit	13.945 mm (0.5490 in)

Master cylinder piston OD – CB750 K '82:
Front	13.957 – 13.984 mm (0.5495 – 0.5506 in)
Service limit	13.945 mm (0.5490 in)

Caliper bore ID – CB750 F '81-82, K, C Type II, 82, SC '82-83:
Front	30.230 – 30.280 mm (1.1902 – 1.1921 in)
Service limit	30.290 mm (1.1925 in)
Rear – F only	27.000 – 27.050 mm (1.0630 – 1.0650 in)
Service limit	27.060 mm (1.0654 in)

Caliper bore ID – CB750 K, C Type I, '82:
Front	30.230 – 30.306 mm (1.1902 – 1.1931 in)
Service limit	30.316 mm (1.1935 in)

Caliper piston OD – CB750 F '81-82 C, K Type II, 82, SC '82-83:
Front	30.148 – 30.198 mm (1.1869 – 1.1889 in)
Service limit	30.140 mm (1.1866 in)
Rear – F only	26.918 – 26.968 mm (1.0598 – 1.0617 in)
Service limit	26.910 mm (1.0594 in)

Caliper piston OD – CB750 C, K Type I, 82:
Front	30.150 – 30.200 mm (1.1870 – 1.1890 in)
Service limit	30.142 mm (1.1867 in)

Note: *During 1981 two types of brake caliper were fitted to the CB750 C and K models. Type I, manufactured by KOKIKO was fitted from August 1st onwards. Type II, made by NISSIN was fitted prior to that date*

Tyre pressures (cold)

	Up to 90 kg (200 lb) load	Above 90 kg (200 lb) load
CB750 K (81-82):		
Front	28 psi (2.00 kg cm^2)	28 psi (2.00 kg cm^2)
Rear	28 psi (2.00 kg cm^2)	40 psi (2.80 kg cm^2)
CB750 C (81-82) SC (82-83):		
Front	32 psi (2.25 kg cm^2)	32 psi (2.25 kg cm^2)
Rear	32 psi (2.25 kg cm^2)	40 psi (2.80 kg cm^2)
CB750 F (81-82):		
Front	28 psi (2.00 kg cm^2)	28 psi (2.00 kg cm^2)
Rear	32 psi (2.25 kg cm^2)	40 psi (2.80 kg cm^2)

Torque settings – 1981 models

Component	kgf m	lbf ft
Front wheel spindle	5.5 – 6.5	40 – 47
Front wheel spindle cap bolts:		
CB750 K, F	1.8 – 2.5	13 – 18
CB750 C (pinch bolt)	1.5 – 2.5	11 – 18
Front brake disc bolts	2.7 – 3.3	20 – 24
Brake hose union bolts	2.5 – 3.5	18 – 25
Front brake caliper bracket bolts	3.0 – 4.0	22 – 29
Front brake caliper bridge bolts CB750 K, C	3.0 – 3.6	22 – 26
Caliper pivot shaft CB750 F	2.5 – 3.0	18 – 22
Caliper mounting bolt CB750 F	2.0 – 2.5	14 – 18
Rear wheel spindle	8.0 – 10.0	58 – 72
Rear wheel sprocket	8.0 – 10.0	58 – 72
Rear brake disc CB750 F	2.7 – 3.3	20 – 24
Rear brake master cylinder CB750 F	3.0 – 4.0	22 – 29
Rear brake torque arm:		
CB750 K, C – front	1.8 – 2.5	13 – 18
CB750 K, C – rear	0.8 – 1.2	6 – 9
CB750 F	1.8 – 2.5	13 – 18

Torque settings – 1982 K, C

Component	kgf m	lbf ft
Brake caliper – Type I:		
Caliper shaft nut	3.0 – 3.6	22 – 26
Caliper mount bolt	1.8 – 2.3	13 – 17
Pad pin retainer bolt	0.5 – 0.8	4 – 6
Brake caliper – Type II:		
Caliper shaft	2.5 – 3.0	18 – 22
Caliper mount bolt	2.0 – 2.5	14 – 18
Pad pin retainer bolt	0.8 – 1.3	6 – 9

Torque settings – 1982-83 SC

Component	kgf m	lbf ft
Brake caliper carrier mounting bolts:		
Right-hand Allen bolt	3.0 – 4.0	22 – 29
Left-hand upper bolt	3.5 – 4.5	25 – 33
Left-hand lower bolt	2.0 – 2.5	14 – 18

Specifications relating to Chapter 6

Fuel gauge sender resistances – SC model

Full	4 – 10 ohms
½ full	28.5 – 36.5
Reserve	58.5 – 80.0

Bulb wattages

Headlamp	60/55W H4 Quartz Halogen

Fuses

Main	30A
Neutral, oil	15A
Headlamp	15A
Turn signal, brake, horn	15A
Tail, instrument, position/parking	15A

Specifications US CB900 F models. Unless shown otherwise all information relates to 1981 and 1982 models

Specifications relating to Chapter 1

Valve timing

Inlet opens at	10° BTDC at 1 mm lift
	63° BTDC at 0 lift
Inlet closes at	35° ABDC at 1 mm lift
	98° ABDC at 0 lift
Exhaust opens at	40° BBDC at 1 mm lift
	70° BBDC at 0 lift
Exhaust closes at	5° ATDC at 1 mm lift
	93° ATDC at 0 lift

Torque settings

Component	kgf m	lbf ft	Notes
Cylinder head cover	1.0	7	
Camshaft bearing caps	1.4	10	
Cylinder head	3.8	27	*
Camshaft sprocket bolts	1.9	14	
Spark plug	1.6	12	
Crankcase 8 mm bolts	2.3	17	*
Alternator rotor bolt	9.0	65	
Primary shaft bolt	9.0	65	
Clutch nut	4.0	29	
Gearbox sprocket	5.0	36	
Connecting rod big-end nuts	3.2	23	
Oil filter centre bolt	3.0	22	
Oil pressure switch	1.8	13	**
Neutral switch	1.8	13	
Sump drain plug	3.8	27	
Oil pump union bolts	2.3	17	
Automatic timing unit	3.5	25	***
Starter clutch bolts	2.8	20	***

Notes: *Apply engine oil to threads and underside of nuts
         ** Apply liquid sealant
         *** Apply Loctite 271 or similar to threads

Specifications relating to Chapter 2

Fuel tank

Total capacity	20 litres (4.4/5.3 Imp/US gal)
Reserve capacity	2.5 litres (0.55/0.66 Imp/US gal)

Carburettors

Make	Keihin
Venturi diameter	30 mm (1.3 in)
Identification No	VB43A (VB43B, Canada)
Float level	15.5 mm (0.61 in)
Primary main jet	68
Secondary main jet	105
Idle speed	1000 ± 100 rpm
Throttle grip free play	2-6 mm ($\frac{1}{8}$ − $\frac{1}{4}$ in)
Fast idle speed	1000 − 2500 rpm
Pilot screw initial opening	$2\frac{1}{2}$ turns out

Specifications relating to Chapter 3

Ignition timing

Retarded	10° BTDC at idle
Advanced	38.5° BTDC at 3100 rpm

Spark plugs

	Cold climate (below 5°C, 41°F)		Standard	
1981 model:				
Make	NGK	ND	NGK	ND
Type	D8EA	X24ES-U	D9EA	X27ES-U
Type (Canada)	DR8ES-L	X24ESR-U	DR8ES	X27ESR-U
1982 model:				
Make	NGK	ND	NGK	ND
Type	DR8ES-L	X24ESR-U	DR8ES	X27ESR-U
Gap	0.6 − 0.7 mm (0.024 − 0.028 in)			

Torque settings

Component	kgf m	lbf ft
Automatic timing unit bolt	3.5	25
Spark plugs	1.6	12

Specifications relating to Chapter 4

Front forks − no specifications available for CB900 F at the time of publication

Rear suspension

Swinging arm bearing type	Needle roller
Suspension unit spring free length − service limit	233 mm (9.17 in)

Torque settings

Component	kgf m	lbf ft
Steering stem nut	10.0	72
Steering top thread nut	1.5	11
Handlebar holder	2.2	16
Fork clamp bolts:		
Upper	1.1	8
Lower	5.0	36
Fork cap bolt	2.3	17
Front wheel spindle clamp bolts	3.5	25
Front wheel spindle nut	6.0	43
Damper rod Allen bolt	2.0	14
Fork drain bolt	0.8	6
Fork air hose union:		
Right	1.8	13
Left	0.6	4
Fork air hose connector	0.6	4
Fork air valve	0.6	4
Rear wheel spindle nut	9.0	65
Rear wheel sprocket	9.0	65
Swinging arm pivot nut	6.5	47
Rear brake torque arm	2.2	16
Rear suspension unit	3.5	25
Engine mounting bolts	4.0	29

Specifications relating to Chapter 5

Brakes

Front	Twin hydraulic disc, twin piston caliper
Rear	Single hydraulic disc, twin piston caliper
Caliper piston OD – service limit:	
Front	30.14 mm (1.187 in)
Rear	26.91 mm (1.059 in)
Caliper bore ID – service limit:	
Front	30.29 mm (1.193 in)
Rear	27.06 mm (1.065 in)

Tyre pressures cold

	Up to 90 kg (200 lb) load	Above 90 kg (200 lb) load
Front	32 psi (2.25 kg cm^2)	32 psi (2.25 kg cm^2)
Rear – 1981 model	32 psi (2.25 kg cm^2)	40 psi (2.80 kg cm^2)
Rear – 1982 model	32 psi (2.25 kg cm^2)	41 psi (2.85 kg cm^2)

Torque settings

Component	kgf m	lbf ft
Brake disc mounting bolts	3.0	22
Brake caliper carrier	3.5	25
Caliper bolt	2.3	17
Caliper pivot bolt	2.8	20

Specifications relating to Chapter 6

Fuses

Main	30A
Neutral, oil	15A
Headlamp	15A
Turn signal, brake, horn	15A
Tail, instrument, position/parking	15A

1 Introduction

This update Chapter covers the later dohc 750 and 900 models produced for the UK and US markets since 1981. To avoid any possible confusion it is useful to summarise the various models as shown below. Where information is available, the applicable engine and frame numbers are indicated, as are the introduction and discontinuation dates. Note that the dates should be treated as a rough guide only; discontinued models are often first sold some months after the date shown. Note that where marked with an asterisk (*) the machine in question is covered by the original text. The remaining models are discussed in more detail later in this Section.

UK 750 models

Model	Engine No.	Frame No.	Dates
K-Z*	RCO1E-2000640 to 2020308	RCO1-2000042 to 2020204	Oct '78 to '80
F-A*	RC04E-2115857 to 2123244	RC04-2116241 to 2123630	Feb '80 to Mar '81
F-B	RC04E-2204207 to 2216772	RC04-2200007 to 2212095	Mar '81 to '82
F2-C	RC04E-2305710 on	RC04-4000150 on	Feb '82 to '83
F-D	RC04E-2400018 on	RC04-2400001 on	Feb '83 to '84

UK 900 models

Model	Engine No.	Frame No.	Dates
F-Z*	SC01E-2000052 to 2015670	SC01-2000042 to 2015598	Jan '79 to Feb '80
F-A	SC01E-2100001 to 2113813	SC01-2100001 to 2113937	Feb '80 to Feb '81
F-B	SC01E-2200046 to 2203256	SC01-22000028 to2203237	Feb '81 to Feb '82
F2-B	SC01E-2206870 to 2225154	SC01-4000342 to 4011049	Mar '81 to Feb '82
F-C	SC01E-2307653 on	SC09-2000017 on	Feb '82 to Feb '83
F2-C	SC01E-2309508 on	SC09-4000424 on	Feb '82 to Mar '83
F-D	SC01E-2400001 on	SC09-4100001 on	Feb '83 to '84
F2-D	SC01E-2401573 on	SC09-4100811 on	Mar '83 to late '83

US 750 models

Model	Engine No.	Frame No.	Introduced
CB750 K:			
1979*	RC01E-2000001 to 2025441	RC01-2000001 to 20252329	Sep '78
1980*	RC01E-2100005 to 2133994	RC01-2100005 to 2116490	Oct '79
1981	RC01E-2200020 to 2242464	RC010* BM200001 to BM209614	Oct '80
1982	RC01E-2300003 on	RC010* CM300003 on	Sep '81
CB750 K (LTD):			
1979*	RC01E-3000001 to 3008840	RC01-3000006 to 3008810	Jan '79
CB750 F (Supersport):			
1979*	RC04E-2000023 to 2010719	RC04-2000023 to 2010669	Jan '79
1980*	RC04E-2100003 to 2117316	RC04-2100003 to 2116089	Oct '79
1981	RC04E-2200005 to 2216705	RC040* BM200001 to BM207766	Oct '80
1982	RC04E-2300001 on	RC040* CM300003 on	Sep '81
CB750 C (Custom):			
1980*	RC01E-2109040 to 2136728	RC01-2200004 to 2220040	Oct '79
1981	RC01E-2200008 to 2242537	RC011* BM100001 to BM125225	Oct '80
1982	RC01E-2300001 on	RC011* CM200003 on	Sept '81
CB750 SC (Nighthawk):			
1982	RC01E-2308884 to 2323081	RC012* CM000019 to CM012103	Jan '82
1983	RC01E-2400041 on	RC012* DM100001 on	Oct '82

US 900 models

Model	Engine No.	Frame No.	Introduced
CB900 F:			
1981	SC01E-2200962 to 2219696	SC010* BM000021 to BM007060	Oct '80
1982	SC01E-2300001 on	SC010* CM100003 on	Sep '81

UK model development

The machines sold in the UK are identified with a model suffix (e.g. F-A, F-B) denoting the model type and year (see above) and this should be quoted together with the engine and frame numbers when ordering spare parts. To assist further with model identification, the main distinguishing features are listed below. Note also that the paintwork colours and decal design changed with each successive model.

CB750 F-B

Introduced in March 1981, the F-B replaced the CB750 F-A model. The main changes from the preceding model were the addition of a vacuum-operated automatic fuel valve mounted above the right-hand carburettor and supplementing the standard manual fuel tap. A new front brake master cylinder was fitted, this having a rectangular lid held by two screws. Less obvious changes include the fitting of an anti-vibration shim between the front brake discs and wheel hub.

CB750 F2-C

The F2-C was equipped with a sports fairing, and this is its most obvious distinguishing feature. Along with the fairing came additional instrumentation (voltmeter and quartz clock) and a revised headlamp. New front brake calipers were fitted, these being of the twin-piston type.

CB750 F-D

The F-D is generally similar to the F2-C, but without the fairing and additional instruments.

CB900 F-A

Basically similar to the earlier F-Z model, the F-A featured new front forks. These are similar in appearance to the earlier design, but are bushed internally and are equipped with linked air caps. Other details include reversed and highlighted Comstar wheels.

CB900 F-B and F2-B

The FB and F2-B models can be distinguished from the F-A by their twin-piston brake calipers and vacuum-operated automatic fuel valve as fitted to the CB750 F-B. The F2-B is the faired version, and also features a revised headlamp, a voltmeter and a quartz clock. Both models have restyled brake discs with an anti-vibration shim fitted between the disc and wheel hub. Less obvious alterations include detail changes to the gearchange mechanism.

CB900 F-C and F2-C

Basically similar to the F-B and F2-B models, the C models have redesigned front brake master cylinders, caliper mounting brackets, discs and wheels. The engine mounting arrangement is revised with rubber-bushed rear mounts controlled by pivoting rods at the front, to reduce the level of vibration transmitted to the frame. The new front forks are fitted with an anti-dive system, whilst the FVQ rear suspension units have remote reservoirs. Other changes include a revised exhaust system with a balance pipe, and redesigned footrests.

CB900 F-D and F2-D

Apart from cosmetic changes, these models are very similar to the "C" versions. Detail changes include minor alterations to some of the gearbox components.

US model development

The detailed model suffix of the UK range is not applied in the case of the US machines, these being identified by a simpler suffix, the model year and in some instances, a model name. In the summary below each model type is described, together with the main identifying features for each year.

CB750 K 1981 – 82

The K model, introduced in 1978, was the base model of the range. Given the popularity of the CB750 C Custom model, the role of the K changed somewhat, and it became more of an economy version. The following features distinguish the later models from the 1978 – 80 machines covered in the earlier Chapters.

The 1981 and 1982 models are equipped with a vacuum-operated automatic fuel valve mounted above the right-hand carburettor and working in conjunction with the manual fuel tap. The front brake master cylinder design was changed, the lid being retained by two, rather than four, retaining screws.

The front forks were modified to include top and bottom bushes to give improved response over the earlier plain type, and linked air caps were fitted. The FVQ rear suspension units were replaced by Showa VHD components. Other changes included a modified headlamp fitted with a renewable quartz-halogen bulb and revised paintwork and graphics. 1982 saw the introduction of a twin piston brake caliper along with a restyled brake disc with slots around the inner edge.

CB750 F Supersport 1981 – 82

The F, or Supersport model was subject to similar changes to those described above for the K model. In the case of the F models, the new forks were of larger diameter (37 mm instead of 35 mm) giving improved rigidity. The Comstar wheels were of a slightly different design and also featured minor changes in the bearing and spacer arrangement.

CB750 C Custom 1981 – 82

The original (1980) Custom model covered in the main text of this manual was little more than a cosmetically-updated K model. Such was its popularity in the US, however, that Honda made all subsequent versions rather more sophisticated, the K being relegated to an economy model role. Amongst the more noticeable differences from 1981 onwards were the new leading-axle front forks, complete with linked air caps. Reversed Comstar wheels were fitted, and the front brake was of the twin disc type, retaining the single-piston calipers of earlier years. The rod-operated rear drum brake was retained. Swinging arm bearings, as on all models from 1981 onward, were of the needle roller type. Rear suspension units were Showa VHD in place of the previous FVQ type. The earlier sealed-beam headlamp was replaced by a new unit with a renewable quartz-halogen H4 bulb. The manual fuel tap was supplemented by a vacuum-operated automatic valve.

CB750 SC Nighthawk 1982 – 83

Another variation on the factory custom theme, the Nighthawk represented a far more integrated styling exercise than the C model. The distinctive bodywork is the major identifying feature of the model, along with cast alloy (as opposed to the usual Comstar) wheels. 37 mm front forks are fitted, complete with linked air caps and TRAC anti-dive with four position adjustment. Also fitted are twin front disc brakes with twin-piston calipers. Other detail changes from its contemporary models are confined to cosmetic modifications which are numerous but which have little material effect in maintenance and repair terms. There are no significant differences between the 1982 and 1983 models.

CB900 F Supersport 1981 – 82

Available in the UK from 1979, the 900 F finally appeared in the US in 1981, providing an alternative to the CB900 C Custom shaft drive model. (The latter model is covered in a separate Haynes Owners Workshop Manual, book number 728). The 900 F combined the larger engine, with chain final drive, with the Supersport chassis of the 750 F. Unlike the 750 version, the 900 featured a rubber-mounted engine to minimise the level of engine vibration transmitted to the frame. Engine movement was controlled by pivoting rods at the front mountings. Bushed, air assisted forks were fitted (but no anti-dive), as were remote reservoir rear suspension units, reversed Comstar wheels, a revised brake master cylinder, twin piston calipers, automatic fuel tap and a quartz halogen headlamp. There are no major differences between the 1981 and 1982 models, apart from the availability of a sports fairing version, presumably equivalent to the UK's CB900 F2-C.

2 Engine: modifications

Removal and installation – CB900 models

1   The engine removal procedure on the later 900 models is slightly changed due to the adoption of anti-vibration engine mountings and a revised exhaust system. The new engine mountings comprise rubber bushed rear mounting points in conjunction with short pivoting links on the front mounting points. This allows the engine to move slightly in the frame cradle, the links controlling the amount of movement and preventing the engine from twisting in the frame.

2   When removing the engine, take off the links at the front upper mounting, marking them so that they can be refitted correctly. The lower front and rear mountings have damper plates fitted; note the position of these and be sure to refit them in the same relative positions during installation. If this precaution is not observed, excessive vibration is likely to result. Check the rubber mountings for wear or damage and renew them if required. Examine the link pivot bushes and renew them if worn. Remove all traces of road dirt and grease the bushes prior to installation.

3   To free the exhaust system, remove the two bolts which secure the system to the sump. Slacken the clamps which retain the inner pair of pipes to the centre section of the system, and the single clamp between the balance pipe halves to the rear of the sump. The exhaust pipe flange nuts and silencer mounting bolts can now be removed and the system lifted away.

4   When refitting the exhaust system, assemble it loosely, leaving 120 mm ($4\frac{3}{4}$ in) between the right-hand and left-hand sump mounting holes. Offer the system up and fit the various fasteners finger tight. Tighten the system down to the torque settings and in the sequence shown overleaf.

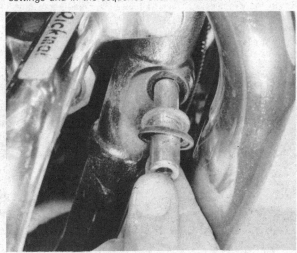

2.2a Clean and lubricate engine mounting pivot and fit into frame lug as shown

Component	kgf m	lbf ft
1 Exhaust port flange nuts	3.3	24
2 Exhaust pipe to sump bolts	5.1	37
3 Silencer mounting bolts	4.0	29
4 Exhaust pipe clamps (3 off)	2.2	16

Valve clearances – 1982 on US models

5   The information given in the Routine Maintenance section of this manual can be followed with the exception of the camshaft alignment marks. On later models two scribed lines will be found in the right-hand end of the exhaust camshaft.

6   Rotate the crankshaft clockwise until index mark 1 aligns with the front section of the gasket face; the exhaust valve clearance for cylinders 1 and 3 can now be checked and recorded. Rotate the crankshaft further clockwise until index mark 2 aligns; at this point the inlet valve clearance for cylinders 1 and 3 can be taken.

7   To check the clearance for cylinders 2 and 4, rotate the crankshaft in a clockwise direction and realign index mark 1 to check the exhaust valve clearances, and then rotate the crankshaft further until the index mark 2 is aligned to check the inlet valve clearances.

3   Fuel line diaphragm unit: examination and renovation

1   The manually-operated fuel tap fitted to the earlier models was supplemented by an in-line diaphragm unit mounted above the right-hand carburettor. When the engine is running, the diaphragm responds to engine vacuum, opening a valve and allowing fuel to flow to the carburettor float bowls. When the engine is stopped and the engine vacuum ceases, the diaphragm moves back to its rest position. The valve closes, blocking the flow of fuel. The addition of the fuel line diaphragm unit obviates the need to turn the fuel tap to the off position each time the machine is parked, and greatly reduces the risk of flooding or leakage if this is forgotten.

2   It should be noted that there is no provision for bypassing the diaphragm unit, and it follows that if the carburettors run dry they cannot easily be primed. In this eventuality, turn on the manual fuel tap having ensured that there is adequate fuel in the tank. Leaving the throttle closed, use the starter to crank the engine for 2 – 3 seconds. Leave the engine to stand for a few moments, then repeat the cranking operation several times. This should succeed in filling the float bowls and the engine should start. If necessary, slacken one of the float bowl drain screws to check that the float bowls have filled.

3   In the event of a suspected fault in the diaphragm unit it should be removed for testing. Start by removing the dual seat and fuel tank, disconnecting the fuel pipe at the tap as the tank is lifted clear. (Remember to turn it off first!.) Disconnect the

fuel outlet pipe at the carburettor and also the vacuum and air vent pipes, leaving them connected to the diaphragm unit. Remove the retaining screws and lift the unit away from the carburettors.

4   To test the unit properly a special vacuum pump/gauge unit is necessary. This is connected to the vacuum pipe and a check should be made to see whether the valve is opened at 10 – 20 mm Hg (0.4 – 0.8 in Hg). For most practical purposes, the test can be approximated as follows. Connect the fuel inlet pipe (from the higher stub at the centre of the unit) to the fuel tap and place the fuel outlet pipe in a suitable container. Turn the fuel tap to the "ON" or "RES" position, then place the vacuum pipe in the mouth and suck. If fuel flows from the outlet pipe the unit can usually be considered serviceable. If the flow of fuel is restricted, have the unit checked by a dealer or check by substituting a new unit.

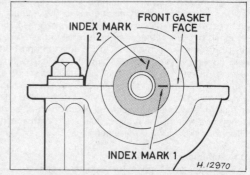

Fig. 7.0 Exhaust camshaft alignment marks for valve clearance check – 1982 on US models
Note: index mark 1 alignment shown

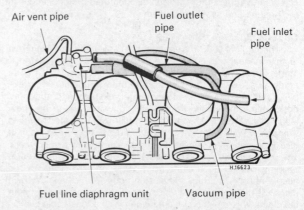

Fig. 7.1 Fuel line diaphragm hose connections

2.2b Do not omit sealing washer on engine mounting bolt

2.2c Fit engine mounting front link and tighten mounting bolts

3.1 Fuel diaphragm unit is mounted on top of right-hand carburettor

4 Ignition amplifier (spark unit): testing – CB900 models 1981 on

1   Refer to Chapter 3, Section 3. noting the following changes to the test procedure.

Test A

2   Trace the pulser leads back to the red block connector and separate it. Turn the ignition on and set the test meter to the 0 – 25 volts dc scale. Connect the positive (+) meter probe to the blue wire (white sleeve) on connector A as in the accompanying line drawing, and the negative (–) probe to a sound earth (ground) point. The meter should show battery voltage (about 12 volts).

Test B

3   With the meter set and connected as described in test A, connect a jumper lead between the blue wire (white sleeve) on the male (spark unit) side of the red connector and a sound earth point. The meter reading should drop to 0 – 2 volts dc.

4   Move the positive (+) meter probe to the yellow wire of connector B. A reading of 12 volts should be indicated.

5   Move the jumper lead from the blue (white sleeve) lead of the red connector to the yellow (white sleeve) lead of the red connector. Voltage should drop to 0 – 2 volts dc.

6   If the readings obtained do not correspond to those shown above, the defective spark unit should be renewed. If in any doubt, have the unit checked by a Honda dealer.

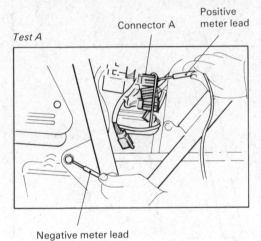

Test A

Connector A

Positive meter lead

Negative meter lead

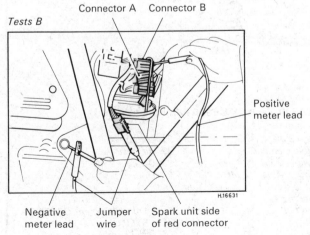

Tests B

Connector A   Connector B

Positive meter lead

Negative meter lead

Jumper wire

Spark unit side of red connector

H.16631

Fig. 7.2 Ignition spark unit test connections – CB900 1981 on models

5 Fork air pressure adjustment

1   The later 750 and 900 models are equipped with air-assisted front forks, the air pressure being variable to provide some measure of spring rate adjustment. The two fork legs are linked by a flexible hose and share a common Schraeder-type valve on the right-hand fork top bolt. This removes the common problem of pressure imbalance where separate valves are fitted.

2   The air pressure should be set according to personal preference within the limits given in the Specifications. Note that it is important not to exceed the maximum pressure figure; if this precaution is ignored, fork operation will suffer and seal damage is likely. To this end, **never** use compressed air from an air line to pressurise the forks. Remember that the total volume inside the forks is limited and it will be impossible to control an air line with sufficient accuracy to avoid the risk of over-pressurisation.

3   Many motorcycle dealers can supply small syringe-type air pumps specifically designed for use on air suspension systems. These are ideal for the purpose and a worthwhile investment. Alternatively, an ordinary bicycle pump will work well, but note that it will be necessary to use a Schraeder-type adaptor. This type of valve is used on car and motorcycle tyres and also on high pressure bicycle tyres and can be purchased from most bike shops. In the author's experience, about 3 – 4 strokes of the pump will be sufficient to bring the forks up to pressure.

4   When checking the pressure it is best to use a pencil-type pressure gauge. These are usually quite accurate and do not lose a lot of pressure during measurement. Despite this last point, bear in mind that a certain amount of pressure loss is unavoidable when checking the pressure. It pays to experiment to find the amount of loss incurred in this way; the fork pressure can then be set so that it falls to the desired setting after the measurement has been made.

5   The recommended air pressure for the CB900 F-A forks is 0 psi although this can be raised to between 7 – 11 psi (0.5 – 0.8 kg/cm^2), especially where a fairing is fitted. Note that on no account must the pressure be raised beyond 14 psi (1.0 kg/cm^2) otherwise serious fork component damage may occur. Later UK CB900 models have a standard recommended air pressure of 11 – 14 psi (0.8 – 1.0 kg/cm^2). The manufacturer gives a maximum air pressure of 42 psi (3.0 kg/cm^2) but it should be noted that at this pressure serious damage to the fork components may occur, and it is strongly advised that the air pressure is kept within the specified range of 11 – 14 psi (0.8 – 1.0 kg/cm^2).

5.1 Fork air pressure can be altered via valve at top of fork leg. Do not omit to refit dust cap

6   Front forks: general description

1   From 1981 onwards a revised type of fork was fitted to all US models and the UK 900 models. In the case of the UK 750 machines the previous unbushed fork was retained. There are detail differences between the forks fitted to the various models in the range, but these relate mainly to internal dimensions, the arrangement of backing washers between the oil seal and top bush, and to specific damper rod seat (oil lock piece) arrangements. These are described in the Specifications and shown in the accompanying line drawings.

2   The most significant change from the earlier types is the adoption of top and bottom bushes. This means that the forks can be built to finer tolerances and thus operate more precisely than the unbushed version. The reduced bearing surface area creates less stiction, allowing the fork to respond more readily to small surface irregularities. From the point of view of maintenance, the use of bushes makes it possible to rebuild a worn fork rather than renew the lower leg or stanchion.

3   On machines fitted with TRAC anti-dive braking systems the lower leg incorporates the anti-dive unit. In response to pressure from the caliper link, the anti-dive unit controls the damping rate of the fork. When the brake is off, the suspension moves normally, but as front brake pressure is transmitted mechanically through a link and pivot bushes to a control valve in the anti-dive case, this increases the compression damping effect, resisting the tendency of the machine to pitch downwards at the front. The degree of anti-dive effect is controlled by an adjuster on the side of the unit, and provision is made to permit normal fork movement when large surface irregularities are encountered.

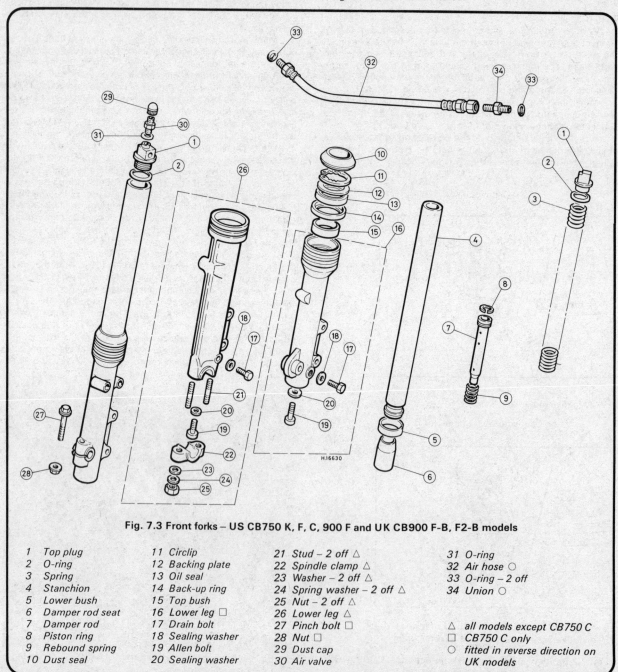

H.16630

Fig. 7.3 Front forks – US CB750 K, F, C, 900 F and UK CB900 F-B, F2-B models

1	Top plug	11	Circlip	21	Stud – 2 off △	31	O-ring
2	O-ring	12	Backing plate	22	Spindle clamp △	32	Air hose ○
3	Spring	13	Oil seal	23	Washer – 2 off △	33	O-ring – 2 off
4	Stanchion	14	Back-up ring	24	Spring washer – 2 off △	34	Union ○
5	Lower bush	15	Top bush	25	Nut – 2 off △		
6	Damper rod seat	16	Lower leg □	26	Lower leg △		
7	Damper rod	17	Drain bolt	27	Pinch bolt □	△	all models except CB750 C
8	Piston ring	18	Sealing washer	28	Nut □	□	CB750 C only
9	Rebound spring	19	Allen bolt	29	Dust cap	○	fitted in reverse direction on
10	Dust seal	20	Sealing washer	30	Air valve		UK models

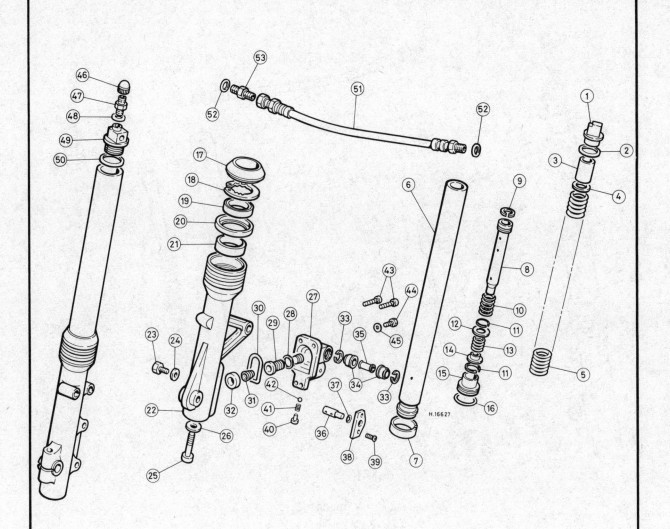

Fig. 7.4 Front forks – US CB750 SC model

1 Top plug	15 Damper rod seat	28 Seal	41 Spring
2 O-ring	16 O-ring	29 Piston	42 Ball
3 Spacer	17 Dust seal	30 O-ring	43 Allen bolt – 4 off
4 Spring seat	18 Circlip	31 Piston spring	44 Drain bolt
5 Spring	19 Oil seal	32 Seal	45 Sealing washer
6 Stanchion	20 Back-up ring	33 Circlip – 2 off	46 Dust cap
7 Lower bush	21 Top bush	34 Boot – 2 off	47 Air valve
8 Damper rod	22 Lower leg	35 Sleeve	48 O-ring
9 Piston ring	23 Drain bolt	36 Selector	49 Top plug
10 Rebound spring	24 Sealing washer	37 O-ring	50 O-ring
11 Circlip – 2 off	25 Allen bolt	38 Selector retaining plate	51 Air hose
12 Washer	26 Sealing washer	39 Screw – 2 off	52 O-ring – 2 off
13 Spring	27 Anti-dive valve housing	40 Detent bolt	53 Union
14 Headed spacer			

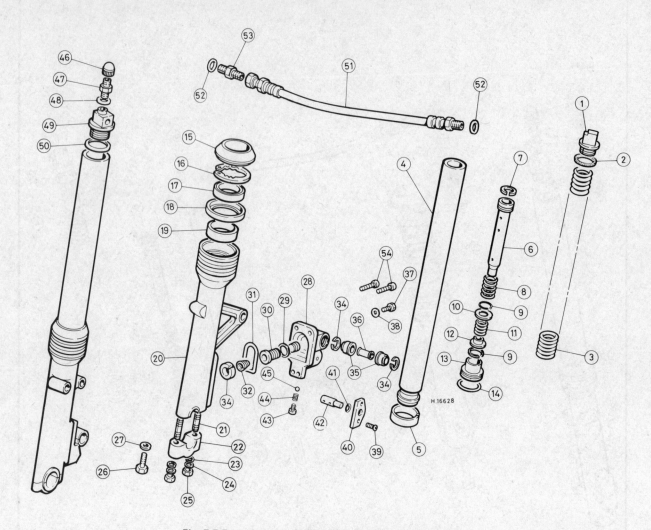

Fig. 7.5 Front forks – UK CB900 F-C, F2-C, F-D and F2-D

1 Top plug	15 Dust seal	29 Seal	42 Selector
2 O-ring	16 Circlip	30 Piston	43 Detent bolt
3 Spring	17 Oil seal	31 O-ring	44 Spring
4 Stanchion	18 Back-up ring	32 Piston spring	45 Ball
5 Lower bush	19 Top bush	33 Seal	46 Dust cap
6 Damper rod	20 Lower leg	34 Circlip – 2 off	47 Air valve
7 Piston ring	21 Stud – 2 off	35 Boot – 2 off	48 O-ring
8 Rebound spring	22 Spindle clamp	36 Sleeve	49 Top plug
9 Circlip – 2 off	23 Washer – 2 off	37 Drain bolt	50 O-ring
10 Washer	24 Spring washer – 2 off	38 Sealing washer	51 Air hose
11 Spring	25 Nut – 2 off	39 Screw – 2 off	52 O-ring – 2 off
12 Headed spacer	26 Allen bolt	40 Selector retaining plate	53 Union
13 Damper rod seat	27 Sealing washer	41 O-ring	54 Screws
14 O-ring	28 Anti-dive valve housing		

7 Front forks: dismantling, renovation and reassembly

1   The general procedure for dealing with most types of forks is broadly similar, and in most respects the details given in Chapter 4, Section 6 can be applied. Note however that the later bushed forks require a slightly revised approach in certain respects, and this is discussed below. This Section relates to all US models from 1981 onwards. In the case of the UK 750 models the original unbushed forks were used and thus this Section does not apply. The UK CB900 F and F2 models from

1981 onwards are covered below.
2   Before the fork legs can be removed it is necessary to disconnect the air hose which connects the two top plugs. Start by removing the dust cap from the valve, then depress the valve core to release air pressure from the forks. Hold the connector hexagon to prevent it turning, then unscrew the air hose union. (The connector is located on the right-hand or left-hand cap bolt according to the model.) Once disconnected, unscrew the remaining end of the hose and remove it.
3   The individual fork legs can now be removed in the usual way, having first released the front wheel, mudguard and the

brake caliper(s). On those machines fitted with a fairing it is desirable, though not absolutely essential, to remove the fairing before attempting to remove the forks. With the fairing removed access is much improved and there is less chance of damaging the fairing surfaces.

4   Clamp the lower leg in a vice by its caliper mounting lugs, using soft jaws to prevent damage. Slacken and remove the cap bolt. This is best done using self-locking pliers – Vise-grips or a Mole wrench for example – with thin strips of hardwood to protect the surface finish of the cap bolt. Lift away the cap bolt, the spacer (where fitted) and the spring. It is important to note the relative position of these components so that they are refitted correctly. Where a variable-rate spring is used the closer-wound coils must be fitted towards the top of the fork, whilst the springs used on the Custom model have their tapered portion at the bottom. Before proceeding further, invert the fork over a drain tray and "pump" it to expel the damping oil.

5   It is worth noting at this stage that Honda recommend a method of removing the fork oil seal without separating the stanchion and lower leg. This requires the removal of the dust seal, circlip and backing plate and the fitting of a tubular drift (part number 07947-3290000). The fork spring and spacer are then removed and the entire fork filled with automatic transmission fluid (ATF). The cap bolt should now be refitted, the air hose hole plugged, and the assembly fitted upright in an hydraulic press. Next, some rag should be wrapped around the seal area and the fork slowly compressed. As the pressure builds in the fork, the seal will be forced up and out of the lower leg.

6   Whilst this method does avoid separating the stanchion and lower leg it does incur a good deal of preliminary work and requires specialised equipment. For this reason it is suggested that this approach is used only where facilities permit. Most owners will be better advised to separate the fork leg, this having the added advantage of allowing a full inspection of the internal components.

7   Before the stanchion and lower leg can be separated, the Allen bolt which secures the damper rod to the lower leg must be unscrewed. If the damper rod rotates inside the lower leg, temporarily refit the spring, spacer and cap bolt to apply pressure to the head of the damper rod.

8   Remove the dust seal from the top of the lower leg by inserting a knife blade or similar between the top surface of the lower leg and the outer flange of the seal. Work the blade round the seal until it can be slid away from its recess. If care is taken, the seal can be removed without damage, but be prepared to renew it if it becomes twisted or torn. Using circlip pliers, remove the circlip which retains the oil seal, backing plate (where fitted) and top bush in the lower leg.

9   The stanchion cannot be withdrawn from the lower leg until the top bush has been displaced. Pull the stanchion outwards until it stops, then push it inwards by an inch or two and pull it sharply outwards to dislodge the bush. Repeat this procedure until the bush is tapped free of the lower leg, then slide the stanchion assembly out of the lower leg.

10   On later models, the damper rod seat, or "oil lock piece" is sealed around its edge by an O-ring and will remain in the bottom of the lower leg unless displaced. It should be removed and the O-ring renewed if any sign of wear or leakage is discovered. To free the damper rod from the stanchion, remove the circlips, washers and spring from its lower end, noting carefully the order in which they are fitted. **Do not** attempt to remove the lower bush from the stanchion; this must not be disturbed unless renewal is required.

11   Examination of the fork components is much the same as has been described for the earlier types, but specific attention should be paid to the bushes. If there are obvious signs of damage, such as scoring of the working surfaces of either bush, they should be renewed. Wear of the stanchion surface is less likely, but if damage is found it too should be renewed. Wear of the bushes can be checked visually, and if the copper-coloured backing material shows through over $\frac{3}{4}$ or more of the surface,

the bushes should be renewed. A worn or damaged bottom bush can be removed after spreading its longitudinal seam with a screwdriver. When fitting the new bush take care not to stretch it any more than is necessary to ease it onto the stanchion.

12   When reassembling the fork leg it will be necessary to find a length of tubing having an internal diameter slightly bigger than that of the stanchion. This can then be used as a tubular drift to seat the top bush in the lower leg. The old bush can be used to bear on the new one during fitting. Lubricate the oil seal with ATF, then press it into position using the tubular drift. Note that the marked face of the seal faces upwards. Refit the backing plate and circlip (radiused edge downwards), then refit the dust seal. When refitting the fork spring and spacer, ensure that they are fitted correctly; if in doubt refer to the accompanying line drawings. Refit the fork leg, mudguard, wheel and brake calipers, paying attention to the torque settings shown in the Specifications. Add the prescribed amount of fork oil to each leg and remember to set the fork air pressure correctly having first pumped the forks up and down a few times to settle the oil level.

7.2 Disconnect connecting hose at unions (arrowed)

7.3a Circlip locates handlebar risers on stanchion and must be removed ...

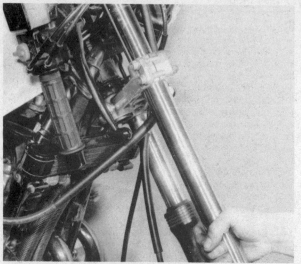

7.3b ... before fork leg is pulled down and clear of yokes

7.4 Where variable rate springs are fitted, close-wound coils must be uppermost

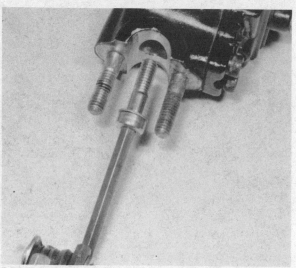

7.7 Unscrew damper rod Allen bolt from base of fork leg

7.8 Free dust seal and remove circlip, then pull stanchion sharply outwards to displace bush and seal

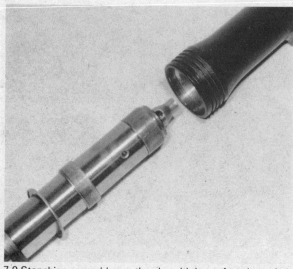

7.9 Stanchion assembly can then be withdrawn from lower leg

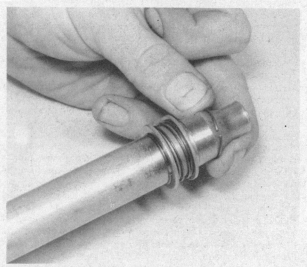

7.10a Prise off the circlip at bottom of damper rod

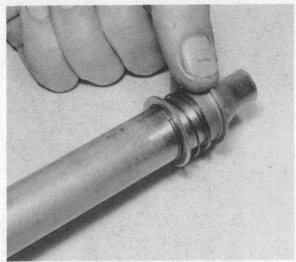

7.10b Remove the headed spacer from inside spring coils ...

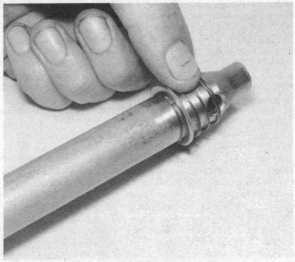

7.10c ... then remove the damper valve spring ...

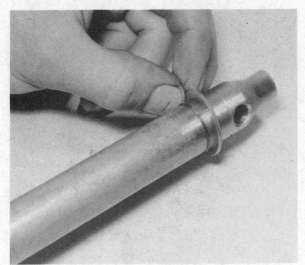

7.10d ... the plain washer and second circlip. Damper rod can then be tipped out of stanchion

7.12a Tap bush fully home, then fit oil seal, backing plate (where fitted) and circlip

7.12b Dust seal fits flush in top of lower leg

8 Anti-dive unit: examination and renovation

1   The anti-dive unit takes the form of a rectangular valve casing bolted to the fork lower leg and connected by a short torque link to the brake caliper. The unit is fitted to one or both of the fork legs, according to the model. Where there are two units fitted, deal with one at a time to avoid interchanging the internal parts. As a general rule it is not necessary to disturb the valve assembly except as part of a fork overhaul, when the valve unit should be removed for inspection and cleaning. If it proves necessary to remove the unit at any other time, note that it will first be necessary to disconnect the caliper at the torque link, release fork air pressure and to drain the fork oil. To facilitate this a drain plug is provided on the valve body. Note that the detent bolt, which passes up into the unit from the underside, should not be disturbed at this stage.

2   Remove the four Allen bolts which retain the unit to the lower leg and lift it away. Referring to the accompanying photographs, remove the detent bolt and shake out the spring and ball, placing them in a container for safe keeping. The seal and piston spring will probably have remained in place in the lower leg and can be lifted out for inspection and cleaning. The

piston should be withdrawn from the valve body. Remove the two screws which retain the selector retainer plate and lift the selector out of the valve body.

3   Check the O-ring seals on the selector and valve piston, and the piston to body seal and valve body seal for wear or damage, renewing them if necessary. Examine the piston surface for scoring, and check its diameter using a vernier caliper. Measure the valve spring free length, comparing the readings obtained with those given in the Specifications.

4   Clean the internal components before reassembly commences, paying particular attention to the orifices in the selector; if these become obstructed, the anti-dive effect will be seriously affected. Lubricate the seals with automatic transmission fluid (ATF) prior to installation. Reassemble by reversing the dismantling sequence, taking care not to damage the seal faces. Refit the unit, tightening the Allen bolts evenly to avoid distortion. Remove and clean the sleeve to which the torque link attaches, lubricating it with silicone grease. Add the recommended quantity of oil to the fork leg and set the fork air pressure correctly before using the machine. Where two units are fitted, check that the same anti-dive setting is selected on both units.

8.1 Anti-dive units also house fork oil drain plug

8.2a Release four mounting bolts and separate anti-dive unit

8.2b From base of unit, remove the screw and sealing washer ...

8.2c ... the damping selector detent spring ...

8.2d ... and the detent ball

8.2e Remove spring and seal from lower leg

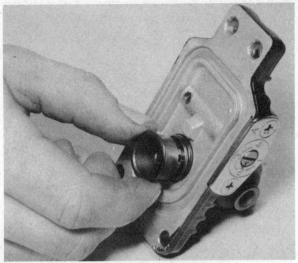

8.2f Withdraw piston from the anti-dive unit

8.2g Release selector retainer ...

8.2h ... and withdraw selector for examination and cleaning

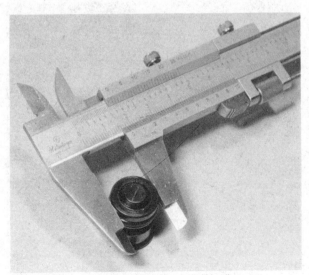

8.3a Measure piston diameter and renew if badly worn

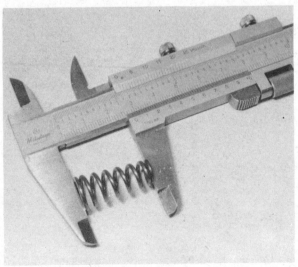

8.3b Check spring free length as shown

8.4 Note cracks around pivot seals – these require renewal

9 Remote reservoir rear suspension units: general

1   The later 900 models make use of revised rear suspension units in place of the earlier FVQ type. The main distinguishing feature is a pressurised remote oil reservoir running parallel to the main damper body, designed to keep the damping oil temperature at a more even level. This in turn reduces the tendency for the damping effect to fall off as the unit heats up in use. For most practical purposes there is little difference between the earlier and late type units in the amount of dismantling that can be undertaken; this is restricted to removal of the spring and is described in Section 11 of Chapter 4. As with the earlier units, a leaking or damaged damper means renewal; it is not possible to rebuild it.

10 Fairing: removal and refitting – F models

1   On models equipped with a fairing it is often either advantageous or necessary to remove it to gain access to certain components or assemblies. A case in point is the cylinder head area and valves; to gain access to the cylinder head cover the fuel tank, fairing and the fairing bracket below the tank must be removed. Fairing removal will obviously add to the time needed for preliminary dismantling, and it is suggested that some time is spent working out a number of jobs that may need to be done while the fairing is off. For example, if valve clearance adjustment is to be undertaken, check whether the machine is also due for a fork overhaul in the near future and if so, carry out both jobs.

2   Start by removing the fairing lowers, or legshields. These are retained at the bottom by two bolts and domed nuts. At the top, prise off the black plastic caps which cover the heads of the two retaining screws and remove them. Note the direction in which the headed collars are fitted through the mounting holes and take care not to lose them. The legshield can now be lifted away and the sequence repeated on the remaining legshield.

3   Locate the fairing wiring harness connectors and separate them. These will be found on the left side of the fairing, just below the storage pocket. The connectors are not easy to reach and separating them will require a fair degree of patience and dexterity. If it proves impossible to reach the connectors it may be easier to wait until the fairing can be lifted partly clear of the machine, but at least one assistant will be needed at this stage.

4   Disconnect the horn leads and remove the horns. Slacken and remove the single bolt on each side which secures the fairing subframe to the fairing rear bracket below the fuel tank. With an assistant supporting the fairing, remove the four 8 mm flange bolts and the two retainers at the front mounting points. Carefully manoeuvre the fairing clear of the machine. Where necessary, remember to disconnect the wiring connectors as they become accessible. Place the fairing on an old blanket or similar to avoid scratching its surface.

5   It is not usually necessary to separate the subframe from the fairing or to remove the fairing wiring or fittings, but if this is done note that the fairing should be reassembled before it is installed on the machine. Installation is largely a matter of reversing the removal sequence. Offer up the fairing and fit the front retainer and rear mounting bolts. Check that the fairing is correctly aligned and not twisted or placed under stress, then tighten the bolts to 2.0 – 3.0 kgf m (15 – 22 lbf ft). Refit the horns and reconnect the horn leads. Reconnect the fairing harness wiring connectors.

6   Refit the legshields and knee pads, again ensuring that they are positioned correctly before tightening the mounting screws. Do not omit the headed collars; these prevent damage to the fairing material. Before using the machine, check that the horn, headlamp and turn signals work normally, and check the headlamp beam alignment.

9.1 Later models are fitted with remote reservoir rear suspension units

10.2a Fairing legshields are retained at the bottom by two bolts

10.2b Prise off the black plastic caps ...

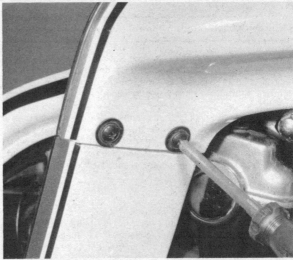
10.2c ... and remove upper mounting screws to free legshields

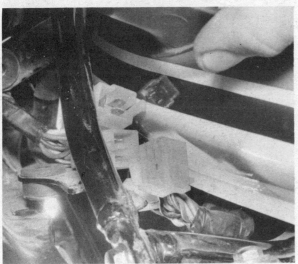
10.3a Separate fairing wiring connectors (centre of picture)

10.3b Another view of connectors – access is limited and will require patience or re-routing during assembly

10.4a Disconnect horn leads and remove from frame

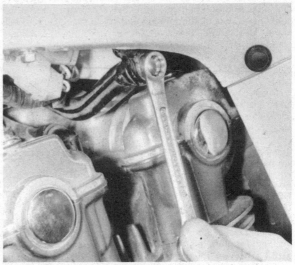
10.4b Release rear mounting flange bolts and retainer clamps on downtubes to free the fairing

11 Front wheel: removal and refitting

1   Whilst the procedure for wheel removal and refitting remains similar to that described in Section 4 of Chapter 5, the revised calipers fitted to the later models necessitate a slightly different approach. In addition, the leading axle type forks used on the C and SC models have a revised wheel spindle arrangement with a pinch bolt at one end, rather than clamps at each end. To avoid confusion, the details relating to specific models are discussed below. Note that on all models it will first be necessary to raise the front wheel clear of the ground, using a jack, wooden blocks or a crate to support the machine below the crankcase.

CB750 C – single piston calipers

2   Remove either the right-hand or left-hand brake caliper, supporting it clear of the wheel and forks. Place a wooden wedge between the brake pads to prevent them from being expelled if the brake lever is accidentally operated. Disconnect the speedometer drive cable at the wheel end. Slacken the pinch bolt at the bottom of the right-hand lower leg, then unscrew and withdraw the wheel spindle, lowering the wheel clear of the forks.

3   To refit the wheel, lift it into position and screw the spindle loosely home. Turn the speedometer drive gearbox anti-clockwise until it stops against the fork lower leg, then tighten the spindle to 5.5 – 6.5 kgf m (40 – 47 lbf ft). Fit the pinch bolt, finger-tight only at this stage. Refit the caliper, tightening the mounting bolts to 3.0 – 4.0 kgf m (22 – 29 lbf ft). Using a 0.7 mm (0.028 in) feeler gauge, check the clearance between the outer face of the right-hand brake disc and the rear edge of the caliper bracket. If the gauge does not fit easily between the two, pull the fork leg outwards, then tighten the pinch bolt to 1.5 – 2.5 kgf m (11 – 18 lbf ft) to secure it. Operate the brake lever several times, then re-check the clearance. Note that if the clearance is inadequate, the brake will tend to drag.

CB750 C, SC – twin piston calipers

4   Remove the right-hand brake caliper, supporting it clear of the wheel and forks. Place a wooden wedge between the brake pads to prevent them from being expelled if the brake lever is accidentally operated. Disconnect the speedometer drive cable at the wheel end by releasing its retaining screw. Slacken the pinch bolt at the bottom of the right-hand lower leg, then unscrew and withdraw the wheel spindle, lowering the wheel clear of the forks.

5   To refit the wheel, fit the speedometer drive gearbox, making sure that it engages correctly in the slots in the wheel hub. Lift the wheel into position and screw the spindle loosely home. Check that the speedometer drive gearbox locates correctly against the lug on the lower leg. Tighten the spindle to 5.5 – 6.5 kgf m (40 – 47 lbf ft). Fit the pinch bolt, finger-tight only at this stage. Refit the caliper, tightening the mounting bolts to 3.0 – 4.0 kgf m (22 – 29 lbf ft). Using a 0.7 mm (0.028 in) feeler gauge, check the clearance between both faces of the brake discs and the caliper bracket. If the gauge does not fit easily between the two, push or pull the fork leg until the clearance is correct. Tighten the pinch bolt to 1.5 – 2.5 kgf m (11 – 18 lbf ft) to secure it. Operate the brake lever several times, then re-check the clearances. Note that if the clearances are inadequate, the brake will tend to drag.

CB750/900 F – twin piston caliper

6   Remove the left-hand brake caliper, supporting it clear of the wheel and forks. Place a wooden wedge between the brake pads to prevent them from being expelled if the brake lever is accidentally operated. Disconnect the speedometer drive cable at the wheel end by releasing its retaining screw. Slacken the four clamp nuts and lower the wheel clear of the forks.

7   To refit the wheel, fit the speedometer drive gearbox, making sure that it engages correctly in the slots in the wheel hub. Position the wheel below the fork legs, then lower the machine until the fork ends rest on the spindle ends. Check that the speedometer drive gearbox locates correctly against the lug on the lower leg. Fit the wheel spindle holders with the arrow marks facing forward and fit the retaining nuts finger tight, starting with the two front nuts. Refit the caliper, tightening the mounting bolts to the prescribed torque figure. Tighten the right-hand holder nuts only, starting with the front nut.

8   Using a 0.7 mm (0.028 in) feeler gauge, check the clearance between both faces of the brake discs and the caliper bracket. If the gauge does not fit easily between the two, push or pull the fork leg until the clearance is correct. Once the clearance is correct, tighten the remaining holder nuts, again starting with the front nut. Operate the brake lever several times, then re-check the clearances. Note that if the clearances are inadequate, the brake will tend to drag.

11.1 Pad pin retainer showing keyhole slots (caliper removed for clarity)

11.2a A: Caliper mounting bolt   B: Caliper pivot bolt. Note model shown has anti-dive, other models similar

11.2b Withdraw pad pins, using pliers where necessary

11.2c Pads can be lifted out of caliper for inspection

11.2d Note direction of fitting of anti-rattle shim

12 Twin piston caliper: general description

1   Later models make use of twin piston brake calipers in place of the single piston type fitted to the earlier models and described in Chapter 5. The new caliper is still of the single sided type in which the caliper body is allowed to slide in relation to the fixed mounting bracket. The pistons work in tandem on elongated brake pads, movement of the caliper body bringing the fixed pad into contact with the disc surface and thus applying equal pressure on both sides. The arrangement should not be confused with opposed piston designs in which the pistons operate from both sides of the caliper, the body of which is fixed in relation to the fork leg and disc.
2   The twin piston design is claimed to offer greater rigidity than earlier designs and to allow a narrower, and thus lighter, disc design. The new caliper is easily identified by the clearly evident paired cylinders. It was fitted to the 750 and 900 F models from 1981 onwards and to the late C and all SC models.

13 Twin piston caliper: pad renewal

Front
1   Remove the pad pin retainer by unscrewing its retaining bolt. This is located inboard of the brake hose union. The retainer has keyhole slots which locate over the ends of the pad pins and can be disengaged and removed once the bolt has been released. Push the caliper body inwards against the mounting bracket. This forces the pistons back into their bores and makes room for the extra thickness of the new pads.
2   Remove the caliper mounting bolt. Note that this is the lower of the two bolts which secure the caliper body to the bracket, and that on machines fitted with anti-dive units it passes through a short torque link. The upper bolt is, in fact, a pivot and need not be removed. Pivot the caliper body upwards and clear of the disc. Using a pair of pointed-nosed pliers, grasp the ends of the pad pins and withdraw them. The pads will now be freed and can be lifted away, as can the anti-rattle shim. When removing the latter, note the direction in which it was fitted to avoid confusion during installation.
3   Clean off any accumulated brake dust from the caliper, taking care not to inhale any of the dust, which has an asbestos content and is thus toxic. Check the caliper carefully for signs of leakage round the pistons; if there are traces of hydraulic fluid which might indicate a leak, trace and rectify the fault before proceeding further. Check that the pistons are fully retracted into their bores. If necessary, they can be pushed inwards using thumb pressure. The pads should be renewed as a matter of course if they are contaminated, badly glazed or scored. Light glazing on otherwise sound pads can be removed by light sanding on coarse abrasive paper. The pad wear limit is denoted by a line in the friction material; renew the pads as a pair if either has worn down to the line.
4   Refit the anti-rattle shim into the caliper ensuring that it locates correctly (see photograph) then fit the new pads. Slide the pad pins into position and refit the pin retainer and its mounting bolt. Swing the caliper down over the disc and fit the mounting bolt, tightening it to its recommended torque setting. After fitting new pads it will be necessary to operate the brake lever repeatedly until the pistons adjust to their new position. When riding the machine, avoid heavy braking for the first 100 miles or so to allow the pads to bed in properly without glazing.

Rear
5   The rear pads can be dealt with in a similar manner to that described above. Note that the dust cover at the rear of the caliper should be removed before the mounting bolt is released and the caliper body swung away from the disc.

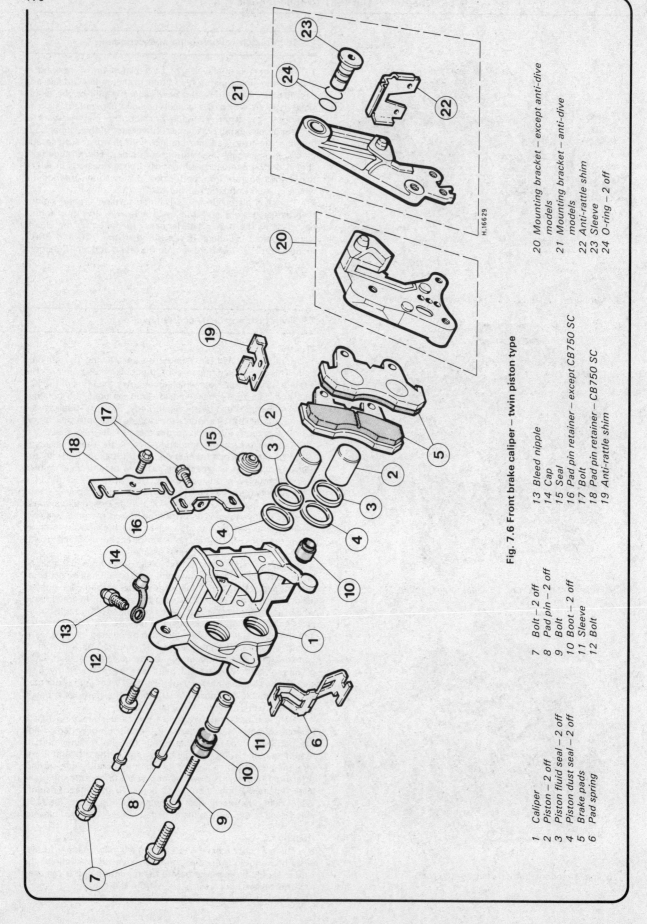

H.16629

Fig. 7.6 Front brake caliper – twin piston type

1 Caliper
2 Piston – 2 off
3 Piston fluid seal – 2 off
4 Piston dust seal – 2 off
5 Brake pads
6 Pad spring

7 Bolt – 2 off
8 Pad pin – 2 off
9 Bolt
10 Boot – 2 off
11 Sleeve
12 Bolt

13 Bleed nipple
14 Cap
15 Seal
16 Pad pin retainer – except CB750 SC
17 Bolt
18 Pad pin retainer – CB750 SC
19 Anti-rattle shim

20 Mounting bracket – except anti-dive models
21 Mounting bracket – anti-dive models
22 Anti-rattle shim
23 Sleeve
24 O-ring – 2 off

14 Twin piston caliper: examination and overhaul

Front

1 The general procedure for dealing with twin piston front calipers is similar to that described in Chapter 5, Section 6, noting the following points. After draining the hydraulic system remove the caliper mounting and pivot bolts and lift the caliper clear of the mounting bracket and disc. Remove the pads and anti-rattle shim as described above. The pistons can be displaced from their bores using compressed air as described in Chapter 5. Note that a strip of wood or similar should be placed in the caliper to prevent one of the pistons from emerging before the other; once both are nearly clear of the bores they can be pulled clear manually. In the absence of compressed air, temporarily reconnect the brake hose and use hydraulic pressure to push the pistons out. With either method, wrap some rag around the caliper to catch any spilt fluid and take care to avoid trapped fingers.

2 Examination and reassembly of the caliper is as described in Chapter 5 with the obvious exception that there are two sets of pistons and bores to be dealt with. It is recommended that the pistons are marked internally with a spirit-based felt marker to ensure that they are refitted in their original bores.

Rear

3 The rear caliper can be dealt with in the same way as the front unit. Note that to gain access to the rear caliper pivot bolt it will first be necessary to release the right-hand rear suspension unit mounting bolt. The unit can then be pushed forwards to gain the required clearance.

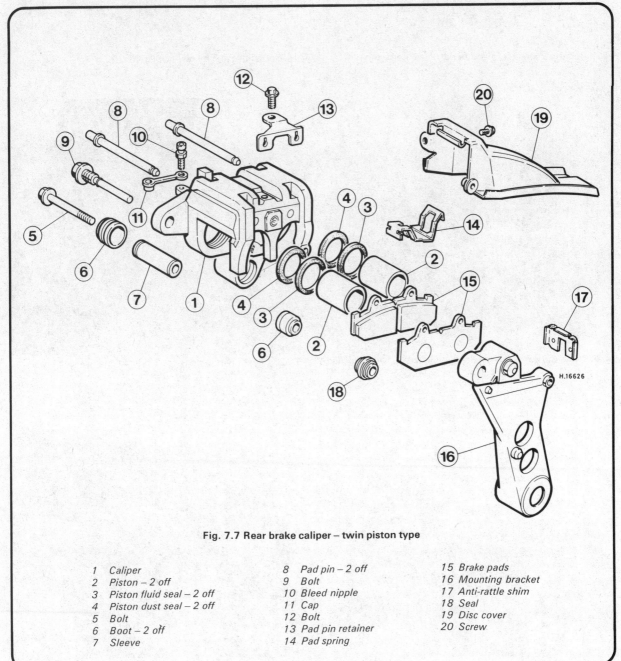

H.16626

Fig. 7.7 Rear brake caliper – twin piston type

1 Caliper	8 Pad pin – 2 off	15 Brake pads
2 Piston – 2 off	9 Bolt	16 Mounting bracket
3 Piston fluid seal – 2 off	10 Bleed nipple	17 Anti-rattle shim
4 Piston dust seal – 2 off	11 Cap	18 Seal
5 Bolt	12 Bolt	19 Disc cover
6 Boot – 2 off	13 Pad pin retainer	20 Screw
7 Sleeve	14 Pad spring	

14.2a Remove pistons to check for scoring or corrosion

14.2b Remove and renew the piston seals ...

14.2c ... and the dust seals, lubricating them with hydraulic fluid

14.2d Check pivot O-rings and lubricate during assembly

14.2e Slide pivot pin into bracket bush. Note thrust shim (arrowed) ...

14.2f ... which butts against ends of pads

15 Regulator/rectifier unit: testing – all models 1981 – 1983

1   To test the regulator/rectifier unit a good quality multimeter will be required. Honda recommend the Sanwa Electrical Tester available as part number 07308 – 0020000 and caution that the use of a cheap or inaccurate meter will give misleading results.

2   Remove the left-hand side panel and disconnect the wiring connectors at the regulator/rectifier unit. Set the meter to the appropriate resistance range (Sanwa SP-10D: Ohms X 1000, Kowa TH-5H: Ohms X 100) then connect the meter probes to the regulator/rectifier terminals as shown in the accompanying chart. If the readings obtained fall significantly outside those shown, renew the unit. Note that the sealed construction of the unit makes repair unpractical.

RECTIFIER   **UNIT : kΩ**

Probe (−) / Probe (+)	Red/White	Green	Yellow 1	Yellow 2	Yellow 3
Red/White		∞	∞	∞	∞
Green	0.5 – 50		0.5 – 50	0.5 – 50	0.5 – 50
Yellow 1	0.5 – 50	∞		∞	∞
Yellow 2	0.5 – 50	∞	∞		∞
Yellow 3	0.5 – 50	∞	∞	∞	

REGULATOR   **UNIT : kΩ**

Probe (−) / Probe (+)	Black	White	Green
Black		1 – 30	0.5 – 20
White	0.5 — 30		1 — 50
Green	0.5 — 20	0.5 – 30	

H.16624

Fig. 7.8 Regulator/rectifier test connections – all models

16 Ignition switch: removal and refitting

1   The later models make use of a revised ignition switch which allows the electrical switch unit to be renewed independently of the lock mechanism. To gain access to the switch it is first necessary to remove the assembly from the top yoke. This in turn requires the removal of the instrument panel assembly.

2   To release the switch unit from the lock body, depress the three plastic tabs which hold it in place; these can be pushed inwards using an electrical screwdriver. With the tabs freed, the switch can be lifted away. When fitting a new switch make sure that all three tabs locate fully in their slots.

17 Headlamp: bulb renewal – US models

1   From 1981 onwards the US models were fitted with European-type headlamp units in place of the sealed-beam units previously fitted. The new unit made use of a 60/55W H4 quartz halogen bulb. This arrangement is the same as that described in Chapter 6, Section 10, for the pre-1981 UK market machines.

18 Headlamp: removal and bulb renewal – all F models equipped with fairing

1   The procedure for gaining access to the headlamp unit is rather more complicated on faired versions of the F model than on the unfaired equivalent; the headlamp unit is housed in a recess in the fairing and is normally covered by a glass shield.

2   To remove the headlamp unit, start by removing the grub screw which retains the headlamp adjuster knob inside the fairing cockpit area, then pull the knob off its spindle. Slacken and remove the large nut, lock washer and plain washer which secure the adjuster. Remove the domed nut and washer adjacent to the adjuster.

3   On the outside of the fairing, remove the glass shield which covers the headlamp unit. This is held in place by a U-section plastic bead, and this must be peeled back to allow the glass to be worked out of position. Take great care not to exert too much pressure on the glass during removal. In cold conditions the plastic bead may prove too stiff to allow the glass to be removed easily. If possible, move the machine into a warm workshop. Failing that, use a hairdryer or a fan heater to warm the plastic bead and soften it. Once the glass is removed the headlamp can be lifted out of the fairing recess and the wiring connector unplugged.

4   Before fitting the glass shield, note that the headlamp horizontal adjuster screw is located in the front edge of the unit and is normally covered by the glass. If necessary, make any alignment adjustment **before** the glass is refitted. Clean any road dirt from the glass shield and the sealing bead. In particular, take care to remove all dirt and fingerprints from the inside face of the glass, while it is accessible. To ease fitting, wipe some soapy water around the bead. With care, the shield can be fitted by hand. Some owners may care to try the method shown in the accompanying photograph. A length of string is threaded around the bead and the glass pressed into place. If the string is now worked around the bead the edge will be lifted over the glass, allowing it to drop into the U-section.

5   Refit the adjuster spindle nut, lock washer and plain washer. Refit the domed nut and washer. Fit the knob, aligning the grub screw hole with the thread in the side of the spindle. Remember to check vertical alignment before using the machine and if necessary adjust it, using the knob inside the fairing cockpit area.

18.2a Remove grub screw which retains headlamp adjuster ...

18.2b ... and pull the knob off adjuster spindle

18.2c Remove adjuster nut and adjacent domed nut to free headlamp

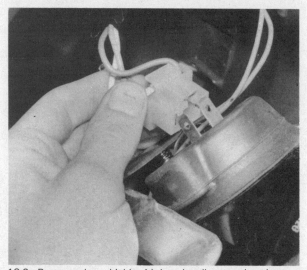

18.3a Remove glass shield, withdraw headlamp and unplug bulb connector ...

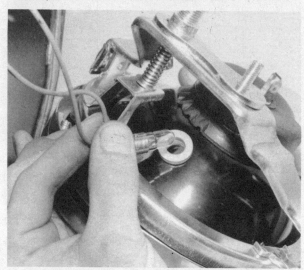

18.3b ... and parking lamp from reflector

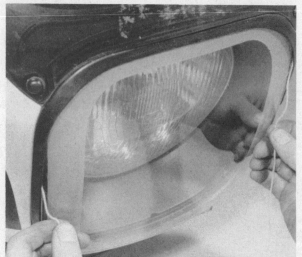

18.4 Use a length of string as shown to work seal around the edge of the glass shield

19   Headlamp: alignment – all F models equipped with fairing

1   Vertical beam adjustment is controlled by a large knob located inside the cockpit area of the fairing, just below the accessory instrument panel. This facilitates adjustment to compensate for passenger or luggage. Horizontal alignment is set using a screw in the front edge of the headlamp, access to which requires the removal of the glass shield in the fairing nose. For details see the previous Section.

20   Accessory instruments: all F models equipped with fairing

1   The faired F models are fitted with a voltmeter and a quartz clock located in an accessory instrument panel just below the fairing screen. The instruments are illuminated internally while the lighting system is on, and the push fit bulbholders can be reached with the instruments in position to facilitate bulb renewal.

2 If either instrument fails to operate, remove it from the panel by releasing the nuts which retain the mounting bracket to the underside of the casing. Disconnect the feed wires (Green and Black in the case of the voltmeter, Green and Red in the case of the clock) and check for battery voltage by connecting

20.2a Instruments are retained by U-bracket and single nut (arrowed)

20.2b Disconnect wiring, lift instrument out of mounting rubber and unplug bulbholder

one of the illuminating bulbs across the two leads. Note that voltage should be present at all times on the clock circuit, but only when the ignition is on in the case of the voltmeter.
3 If the fault does not lie in the wiring, it will be necessary to renew the instrument. Note that although the instruments are of standard automotive size, car instruments may not be able to resist water and vibration as well as the original types. For this reason it is preferable to fit original Honda replacement units.

21 Fuel gauge system: testing

1 The CB750 SC model is equipped with a fuel gauge controlled by a float operated variable resistance, known as a sensor or sender unit, mounted inside the fuel tank. If the accuracy of the gauge is suspect, the sender resistances at various fuel levels should be checked as described below.
2 Switch off the ignition and remove the key for safety. Place the machine on its centre stand on level ground and drain completely the fuel tank, taking the normal precautions to avoid any fire risk. Locate and separate the two-pin connector (Yellow/white and Green leads) below the front edge of the fuel tank. Set a multimeter to the 0-100 ohms scale and connect the test probes to the Yellow/white and Green leads on the sender unit side of the connector.
3 Measure the resistance of the sender unit at the reserve, half and full positions by adding the amount of fuel shown below.

Position	Litres	US Gallons	Imp Gallons	Resistance
Reserve	4.8-6.9	1.27-1.82	1.06-1.52	58.5-80.0 ohms
Half	7.1-10.7	1.88-2.83	1.56-2.35	28.5-36.5 ohms
Full	12.9-15.9	3.41-4.20	2.84-3.50	4.0-10.0 ohms

If the readings obtained correspond with those shown above, the sender unit can be considered serviceable; this means that the fault must lie in the instrument head. The gauge unit is an integral part of the tachometer and if renewal is necessary it must be replaced as an assembly.
4 If the sensor unit readings differed significantly from those shown, drain the fuel tank and remove it. Remove the four nuts which retain the sender unit to the base of the tank and lift it away, taking care not to twist or bend the float arm. Check that the float arm moves smoothly up and down with no signs of sticking. Measure the sender resistances at the FULL (fully raised) and EMPTY (fully down) positions. If these do not correspond with those shown below the sender must be considered faulty and renewed. When fitting the sender, make sure that the O-ring seal is in good condition and check for leakage before refitting the tank.

Sender unit resistances:
    Full 1-6 ohms
    Empty 103-117 ohms

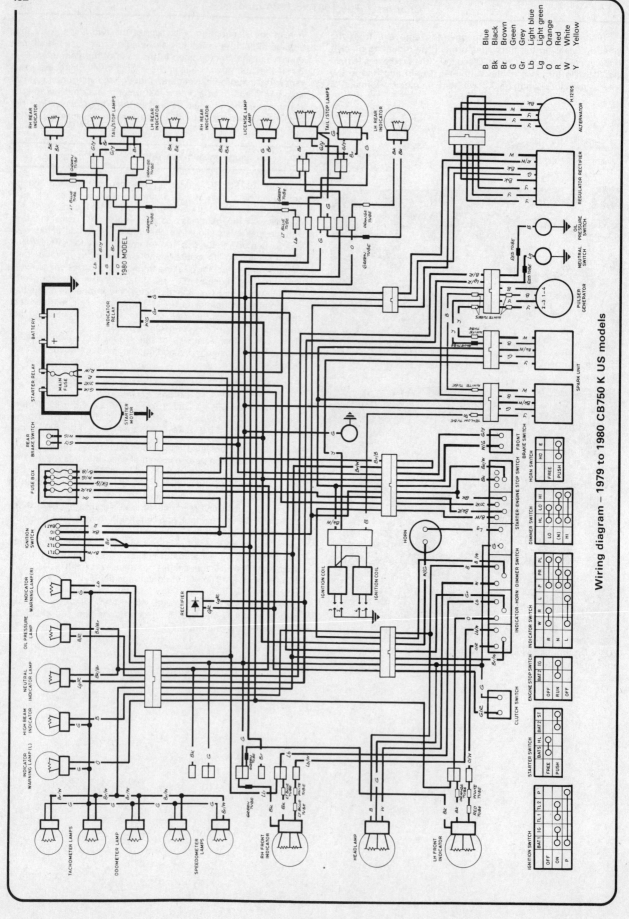

Wiring diagram – 1979 to 1980 CB750 K US models

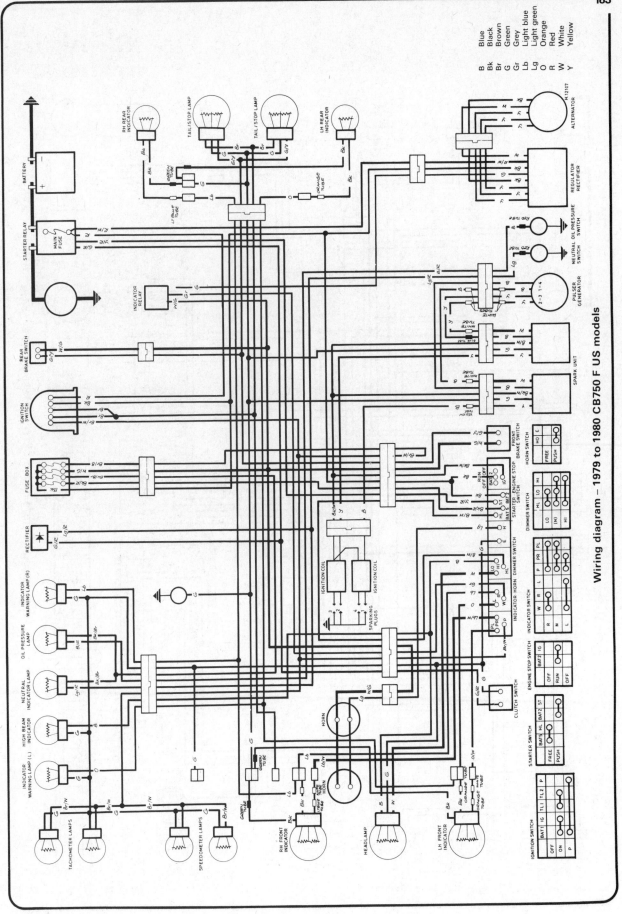

Wiring diagram – 1979 to 1980 CB750 F US models

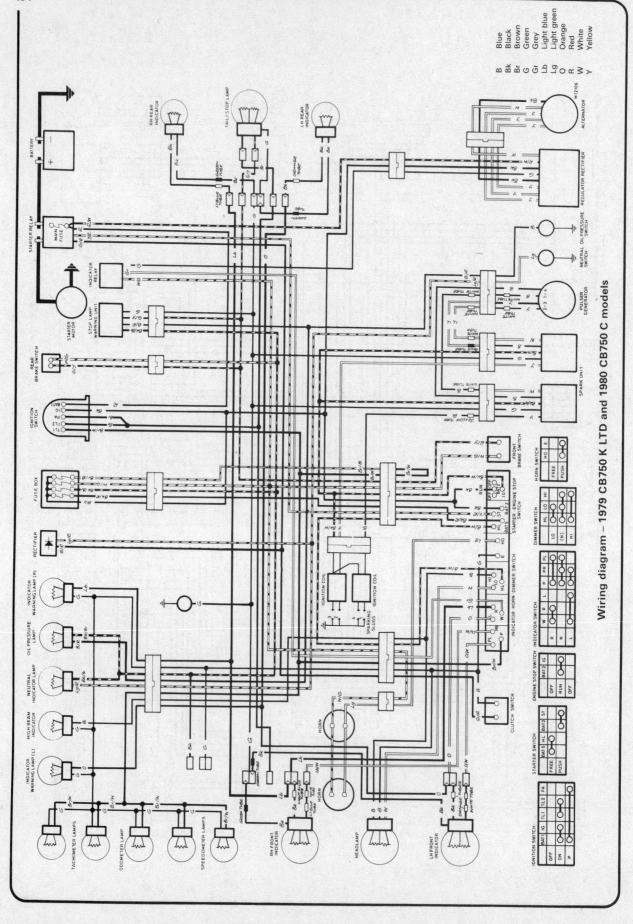

Wiring diagram – 1979 CB750 K LTD and 1980 CB750 C models

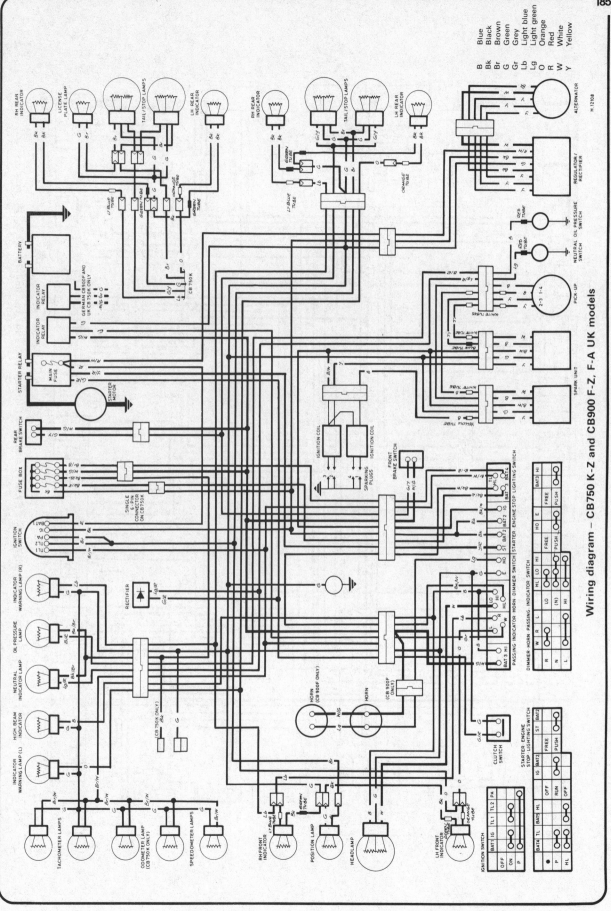

Wiring diagram – CB750 K-Z and CB900 F-Z, F-A UK models

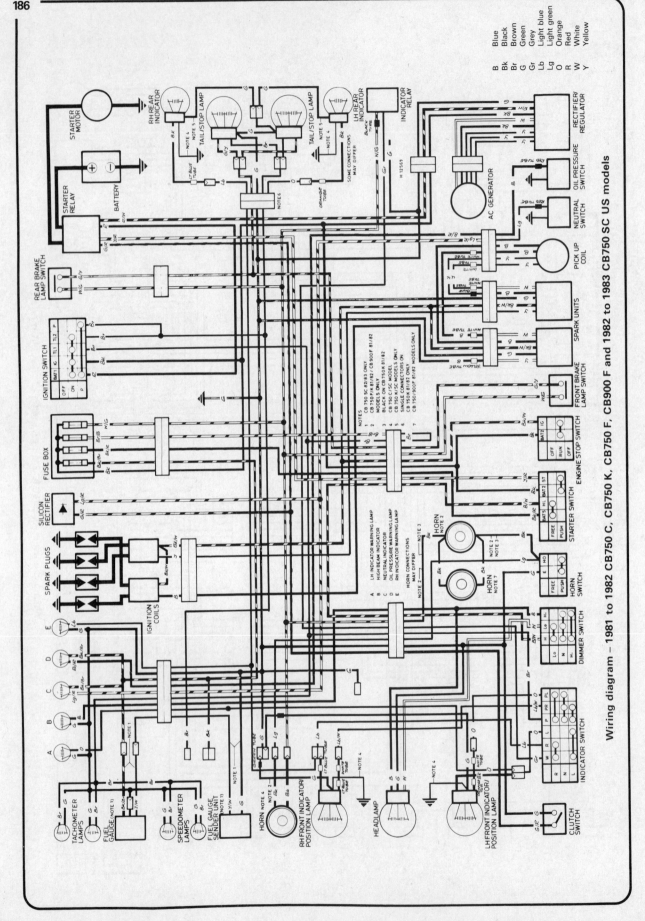

Wiring diagram – 1981 to 1982 CB750 C, CB750 K, CB750 F, CB900 F and 1982 to 1983 CB750 SC US models

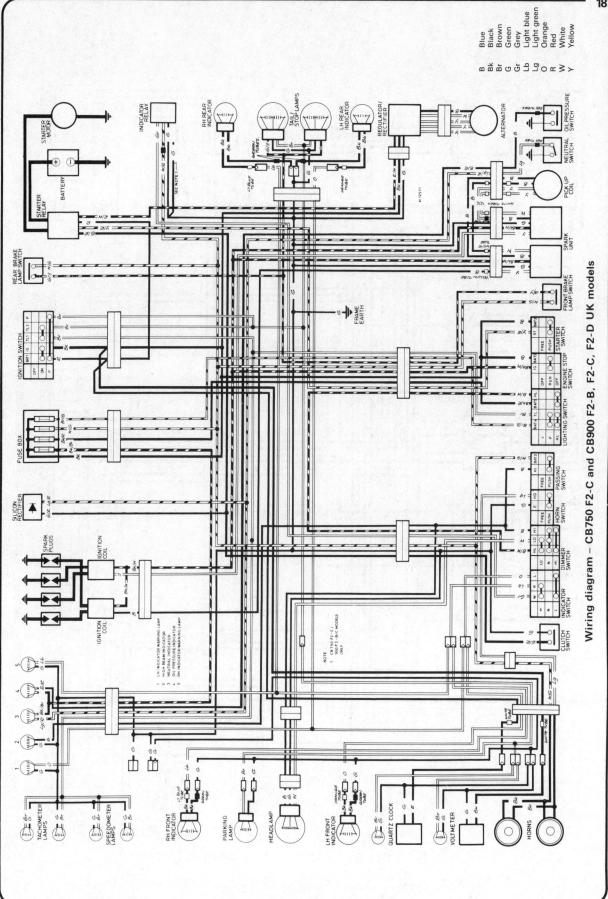

Wiring diagram – CB750 F2-C and CB900 F2-B, F2-C, F2-D UK models

Colour code		
B	Blue	
Bk	Black	
Br	Brown	
G	Green	
Gr	Grey	
Lb	Light blue	
Lg	Light green	
O	Orange	
R	Red	
W	White	
Y	Yellow	

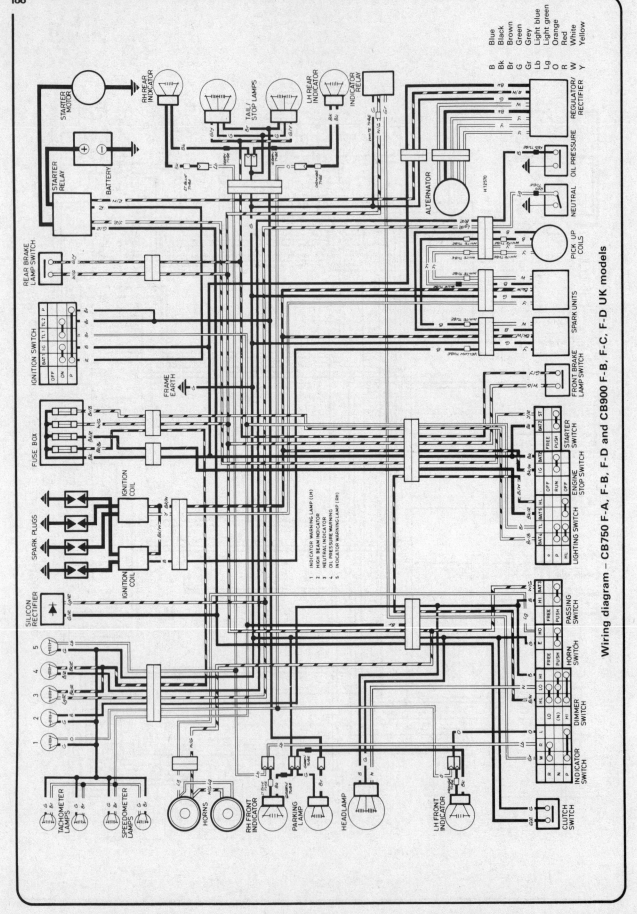

Wiring diagram – CB750 F-A, F-B, F-C, F-D and CB900 F-B, F-D UK models

English/American terminology

Because this book has been written in England, British English component names, phrases and spellings have been used throughout. American English usage is quite often different and whereas normally no confusion should occur, a list of equivalent terminology is given below.

English	American	English	American
Air filter	Air cleaner	Number plate	License plate
Alignment (headlamp)	Aim	Output or layshaft	Countershaft
Allen screw/key	Socket screw/wrench	Panniers	Side cases
Anticlockwise	Counterclockwise	Paraffin	Kerosene
Bottom/top gear	Low/high gear	Petrol	Gasoline
Bottom/top yoke	Bottom/top triple clamp	Petrol/fuel tank	Gas tank
Bush	Bushing	Pinking	Pinging
Carburettor	Carburetor	Rear suspension unit	Rear shock absorber
Catch	Latch	Rocker cover	Valve cover
Circlip	Snap ring	Selector	Shifter
Clutch drum	Clutch housing	Self-locking pliers	Vise-grips
Dip switch	Dimmer switch	Side or parking lamp	Parking or auxiliary light
Disulphide	Disulfide	Side or prop stand	Kick stand
Dynamo	DC generator	Silencer	Muffler
Earth	Ground	Spanner	Wrench
End float	End play	Split pin	Cotter pin
Engineer's blue	Machinist's dye	Stanchion	Tube
Exhaust pipe	Header	Sulphuric	Sulfuric
Fault diagnosis	Trouble shooting	Sump	Oil pan
Float chamber	Float bowl	Swinging arm	Swingarm
Footrest	Footpeg	Tab washer	Lock washer
Fuel/petrol tap	Petcock	Top box	Trunk
Gaiter	Boot	Torch	Flashlight
Gearbox	Transmission	Two/four stroke	Two/four cycle
Gearchange	Shift	Tyre	Tire
Gudgeon pin	Wrist/piston pin	Valve collar	Valve retainer
Indicator	Turn signal	Valve collets	Valve cotters
Inlet	Intake	Vice	Vise
Input shaft or mainshaft	Mainshaft	Wheel spindle	Axle
Kickstart	Kickstarter	White spirit	Stoddard solvent
Lower leg	Slider	Windscreen	Windshield
Mudguard	Fender		

Index